Deepti R. Bathula · Anoop Benet Nirmala ·
Nicha C. Dvornek · Sindhuja T. Govindarajan ·
Mohamad Habes · Vinod Kumar · Ahmed Nebli ·
Thomas Wolfers · Yiming Xiao
Editors

Machine Learning in Clinical Neuroimaging

7th International Workshop, MLCN 2024
Held in Conjunction with MICCAI 2024
Marrakesh, Morocco, October 10, 2024
Proceedings

Editors
Deepti R. Bathula (iD)
Indian Institute of Technology Ropar
Rupnagar, Punjab, India

Anoop Benet Nirmala (iD)
Manipal Institute of Technology
Manipal, Karnataka, India

Nicha C. Dvornek (iD)
Yale University
New Haven, CT, USA

Sindhuja T. Govindarajan (iD)
University of Pennsylvania
Philadelphia, PA, USA

Mohamad Habes (iD)
University of Texas Health Science Center at
San Antonio
San Antonio, TX, USA

Vinod Kumar (iD)
Max Planck Institute for Biological
Cybernetics
Tübingen, Germany

Ahmed Nebli (iD)
Heinrich Heine University Düsseldorf
Düsseldorf, Germany

Thomas Wolfers (iD)
The University of Tübingen
Tübingen, Germany

Yiming Xiao (iD)
Concordia University
Montréal, QC, Canada

ISSN 0302-9743 ISSN 1611-3349 (electronic)
Lecture Notes in Computer Science
ISBN 978-3-031-78760-7 ISBN 978-3-031-78761-4 (eBook)
https://doi.org/10.1007/978-3-031-78761-4

This Springer imprint is published by the registered company Springer Nature Switzerland AG
The registered company address is: Gewerbestrasse 11, 6330 Cham, Switzerland

If disposing of this product, please recycle the paper.

Lecture Notes in Computer Science 15266

Founding Editors

Gerhard Goos
Juris Hartmanis

Editorial Board Members

The series Lecture Notes in Computer Science (LNCS), including its subseries Lecture Notes in Artificial Intelligence (LNAI) and Lecture Notes in Bioinformatics (LNBI), has established itself as a medium for the publication of new developments in computer science and information technology research, teaching, and education.

LNCS enjoys close cooperation with the computer science R & D community, the series counts many renowned academics among its volume editors and paper authors, and collaborates with prestigious societies. Its mission is to serve this international community by providing an invaluable service, mainly focused on the publication of conference and workshop proceedings and postproceedings. LNCS commenced publication in 1973.

Preface

The field of neuroimaging research is undergoing rapid evolution, characterized by the emergence of novel computational techniques and the increasing availability of large publicly available datasets. As computational power and algorithmic sophistication increase, impressive strides have been made to extract meaningful information from complex brain imaging data. However, translating these cutting-edge algorithms into clinical practice has been slower than anticipated, driven by challenges related to reproducibility, generalizability, and clinical utility. Addressing these challenges requires collaboration among engineers, clinicians, and neuroimaging experts. Our workshop aimed to encourage interdisciplinary dialogue between experts in machine learning and clinical neuroimaging.

The 7th International Workshop on Machine Learning in Clinical Neuroimaging (MLCN 2024) was held on October 10, 2024, as a satellite event of the 27th International Conference on Medical Image Computing and Computer-Assisted Intervention (MICCAI 2024) in Marrakesh, Morocco. The call for papers was made on May 14, 2024, and the submission window closed on June 29, 2024. Three or more program committee members reviewed each of the 28 submitted manuscripts in a double-blinded review process. The 16 accepted papers were presented and discussed by the authors at the MLCN workshop and contained methodologically sound, thematically fitting, and novel contributions to the field of clinical neuroimaging.

The accepted papers showcase significant advancements in neuroimaging research, particularly in the analysis of quantitative, structural, and functional MRI and EEG time-series data. They are covered under the sections "Machine Learning" and "Clinical Applications", although all papers carry clinical relevance and provide methodological novelty. These papers heavily feature various deep learning approaches, including graph, diffusion, and transformer-based models, highlighting the fusion of advanced computational techniques with clinical neuroimaging data. The clinical significance of these studies lies in their potential to enhance diagnostic precision and improve patient outcomes for neurodevelopmental conditions such as autism spectrum disorder (ASD) and attention deficit hyperactivity disorder (ADHD), as well as neurological conditions such as Parkinson's disease, Alzheimer's disease, and stroke. Reproducibility is a key focus across these studies, with all models developed being available open source. This transparency ensures that other researchers can validate and expand upon these findings, promoting further innovation in the field.

This workshop was organized once again through the collective efforts of authors, organizing committee members, and workshop participants. We extend our gratitude to all the speakers and participants for their significant contributions, which played a crucial role in the success of the MLCN 2024 workshop.

Deepti R. Bathula
Anoop Benet Nirmala
Nicha C. Dvornek
Sindhuja T. Govindarajan
Mohamad Habes
Vinod Kumar
Ahmed Nebli
Thomas Wolfers
Yiming Xiao

Organization

Steering Committee

Christos Davatzikos — University of Pennsylvania, USA
Seyed Mostafa Kia — University of Tilburg, The Netherlands
Andre Marquand — Donders Institute, The Netherlands
Jonas Richiardi — Lausanne University Hospital, Switzerland

Organizing Committee/Program Committee Chairs

Deepti R. Bathula — Indian Institute of Technology Ropar, India
Anoop Benet Nirmala — Manipal Institute of Technology, India
Nicha C. Dvornek — Yale University, USA
Sindhuja T. Govindarajan — University of Pennsylvania, USA
Mohamad Habes — University of Texas Health Science Center at San Antonio, USA
Vinod Kumar — Max Planck Institute for Biological Cybernetics, Germany
Ahmed Nebli — Heinrich Heine University Düsseldorf, Germany
Thomas Wolfers — University Clinic Tübingen, Germany
Yiming Xiao — Concordia University, Canada

Program Committee/Reviewers

Mumu Aktar — Concordia University, Canada
Mathilde Antoniades — University of Pennsylvania, USA
Anoop Benet Nirmala — University of Texas Health Science Center at San Antonio, USA
Fabian Bongratz — Technical University of Munich, Germany
Matías N. Bossa — Vrije Universiteit Brussel, Belgium
Zhuotong Cai — Yale University, USA
Owen T. Carmichael — Pennington Biomedical Research Center, USA
Yi Hao Chan — Nanyang Technological University, Singapore
Tara Chand — Jindal Institute of Behavioural Sciences, India
Kai-Cheng Chuang — Duke University, USA
Yuhan Cui — University of Pennsylvania, USA
Niharika S. D'Souza — IBM Research, USA
Nikhil J. Dhinagar — University of Southern California, USA

Contents

Clinical Applications

Machine Learning

Parkinson's Disease Detection from Resting State EEG Using Multi-head Graph Structure Learning with Gradient Weighted Graph Attention Explanations

Christopher Neves[1]([✉])(iD), Yong Zeng[2](iD), and Yiming Xiao[1](iD)

[1] Department of Computer Science and Software Engineering, Concordia University,
Montreal, QC, Canada
christopher.neves@hotmail.ca, yiming.xiao@concordia.ca
[2] Concordia Institute for Information Systems Engineering, Concordia University,
Montreal, QC, Canada
yong.zeng@concordia.ca

Abstract. Parkinson's disease (PD) is a debilitating neurodegenerative disease that has severe impacts on an individual's quality of life. Compared with structural and functional MRI-based biomarkers for the disease, electroencephalography (EEG) can provide more accessible alternatives for clinical insights. While deep learning (DL) techniques have provided excellent outcomes, many techniques fail to model spatial information and dynamic brain connectivity, and face challenges in robust feature learning, limited data sizes, and poor explainability. To address these issues, we proposed a novel graph neural network (GNN) technique for explainable PD detection using resting state EEG. Specifically, we employ structured global convolutions with contrastive learning to better model complex features with limited data, a novel multi-head graph structure learner to capture the non-Euclidean structure of EEG data, and a head-wise gradient-weighted graph attention explainer to offer neural connectivity insights. We developed and evaluated our method using the UC San Diego Parkinson's disease EEG dataset, and achieved 69.40% detection accuracy in subject-wise leave-one-out cross-validation while generating intuitive explanations for the learnt graph topology.

Keywords: EEG · Graph Neural Network · Contrastive Learning · Explainable AI · Parkinson's disease

1 Introduction

Parkinson's Disease (PD) is the second most common neurodegenerative disorder worldwide [24]. Primarily characterized by motor symptoms, the complex disease can also include psychiatric and cognitive issues. MRI-based biomarkers

© The Author(s), under exclusive license to Springer Nature Switzerland AG 2025
D. R. Bathula et al. (Eds.): MLCN 2024, LNCS 15266, pp. 3–12, 2025.
https://doi.org/10.1007/978-3-031-78761-4_1

have attracted major attention, including biochemical alteration shown in quantitative MRI and structural/functional connectivity changes revealed by diffusion and functional MRI [17]. However, electroencephalography (EEG), which records electric signals from a network of locations on the scalp is a much more cost-effective neuroimaging tool with higher temporal resolution than MRI that has also been investigated to provide neurological insights and potential biomarkers for the disease. This is especially true for remote or less privileged regions, where MRI scanners are difficult to access.

Recently, deep learning (DL)-based techniques have provided excellent outcomes for EEG analysis, but several challenges remain. **First**, most existing DL techniques for EEG rely on Convolutional Neural Networks (CNNs) that aggregate signals across channels [7,12], but such approaches can miss key spatial characteristics of EEG signals, limiting clinically relevant brain connectivity insights and explainability. **Second**, to better incorporate spatial information, graph neural networks (GNNs) that model different EEG sensors and their relationships as nodes and edges of a graph (often represented as an adjacency matrix) have been proposed. However, although stationary connectivity metrics, such as the Pearson Correlation Coefficient (PCC) or Absolute Cross-Correlation (ACC) are straightforward for deriving the graph for GNN, they often fail to capture non-stationary connectivity, overestimate the correlation between adjacent nodes due to mixing of electrical signals over the scalp surface, and may not provide true functional connectivity insights in many situations. **Third**, EEG data sampled at high frequencies often involves very long sequences, which can pose challenges for commonly used sequential DL models to capture task-relevant features. Recently, Li *et al.* [14] tackled this issue with an effective convolutional model called Structured Global Convolution (SGConv) that has surpassed state-of-the-art sequence models, including Transformers [25] and Structured State Spaces [8], by designing a global convolutional kernel that can span the length of the entire sequence. **Finally**, compared with other medical imaging data, the typically small cohort sizes of EEG datasets can pose challenges for developing robust DL techniques in the domain.

In this work, we aim to address the aforementioned issues with three contributions. **First**, we combined structured global convolutions [14] and self-supervised contrastive learning to better model complex and long sequences of EEG data with a limited cohort for the first time; **Second**, we proposed a novel dynamic multi-head graph structure learning technique to learn the adjacency matrix of the underlying EEG data without imposing potential biases in contrast to conventional static GNN methods; **Third**, to enhance explainability of our DL model for potential clinical insights, we introduced a new technique based on head-wise gradient-weighted attentions to generate an informative adjacency matrix to reveal key task-relevant connectivities in the learnt graph. The proposed method is demonstrated for PD detection with resting state EEG.

2   Related Works

To date, several GNN-based methods [11] have been explored for EEG analysis, particularly for seizure detection in epilepsy. Traditionally, manually defined EEG features, such as Short Time Fourier Transform [4], power spectral density [10], and selective frequency bands [21] have been used in machine/deep learning, but can introduce biases while being time-consuming and expertise-demanding. Therefore, automatic feature extraction methods have become more desirable to reduce biases and improve efficiency. Among these, Dissanayake *et al.* [6] and Sun *et al.* [22] used stacked Long Short-Term Memory (LSTM) networks and Transformers to generate feature embeddings. Li *et al.* [14] proposed the Structural Global Convolution, which showed superior ability to model long and complex sequential signals than prior approaches. Using EEG feature embeddings as node features, different GNN designs incorporating temporal features and spatial properties of EEG data have been devised. One notable trend is the rise of attention-based GNNs, which allow for the visualization of salient edges relevant to the designated tasks to enhance DL model transparency. He *et al.* [9] used a graph attention network (GAT) in conjunction with a bi-directional LSTM for seizure detection and Demir *et al.* [5] used a GAT with additional temporal convolutions to decode motor signals. To mitigate issues with static graphs, Tang *et al.* [23] and Song *et al.* [21] employed the concept of attention to learn the graph adjacency matrix instead of the attention weights between nodes (as in GATs). However, both of their formulations use a single attention head. In EEG-based PD analysis, Chang *et al.* [2] developed a GNN that learns attention coefficients with a graph sparsity constraint to modulate the node feature vectors for PD detection during an auditory oddball task. Further explorations are still required to enhance the efficiency, accuracy, robustness, and transparency of DL-based EEG analysis, especially for GNN-based approaches.

3   Methods and Materials

Figure 1 outlines an overview our proposed DL architecture, which is composed of a feature encoder (LongConv feature encoder), a multi-head graph structure learner (MH-GSL), a Chebyshev GNN, and a classifier made of fully connected layers for PD vs. Healthy classification.

3.1   Feature Encoder with Contrastive Learning

Following the success of Structured Global Convolutions (SGConv) [14] for modeling long sequential data in deep learning tasks, we incorporate it into our EEG feature encoder design, which encodes the input EEG signal to $\hat{X}_e \in \mathbb{R}^{C \times d_m}$ (C is the number of channels and d_m is model dimension). Specifically, we follow the feature extraction network setup in the work of Vetter *et al.* [27], who modify the Structured Global Convolution layer from its original formulation to have more fine-grained control over its kernel size (referred to as SLConv in Fig. 1).

The feature extraction network (called LongConv) consists of interleaved masked 1D convolutions, which project the input channels to a set of hidden ones while SLConv layers extract long-range temporal information from each hidden channel. Each masked 1D convolution is followed by a batch normalization layer and a GELU activation. In our adapted LongConv feature encoder design, we add an additional max pooling operation followed by a 1D convolution (Conv1D) to their network structure before the MH-GSL and Chebyshev GNN layers. To alleviate some of the issues presented by the large inter-subject variability of EEG and the relatively small dataset size, we pretrained the LongConv encoder using the SimCLR [3] framework. First proposed for natural images, SimCLR learns self-supervised data representation by maximizing agreement between differently augmented versions of the same data sample based on a contrastive loss in the latent space. For EEG contrastive learning (CL), we adopted the data augmentations by Mohsenvand *et al.* [16], including combinations of random additive Gaussian noise, random signal masking, a flip along either the signal or electrode dimension or random DC shifts. During training, we used a simple two-layer feed forward network as the projector after the LongConv encoder to obtain a latent space representation used to compute the InfoNCE loss [18]. We used a learning rate of 0.0001, a temperature of 0.005 [16], and a batch size of 100 over 160 SimCLR training epochs.

3.2 Multi-head Graph Structure Learner

Graph topology of EEG signals obtained from stationary connectivity measures and/or the physical distance between electrodes for GNN learning can be misleading and sub-optimal. To tackle this, we proposed a novel graph structure learner (GSL) using multi-head attention. Based on the graph structure layer by Tang *et al.* [23], which adopts the self-attention mechanism [25] to learn edge weights, we extended this approach to include multiple attention heads. Thus, the resulting graph structure learner can attend to different graph representations (adjacency matrices) in parallel, with each attention head providing the edge weights for its paired graph representation. Then, each head-wise learnt graph representation, together with the encoded EEG features are passed to a Chebyshev GNN, updating the features with the learnt spatial relationships. The output of the Chebyshev GNN for each head is then concatenated and projected back to the model dimension d_m using a linear layer. The adjacency matrix $A_h \in \mathbb{R}^{C \times C}$ for a single attention head h out of H heads is given by:

$$
\begin{aligned}
Q_h = \tilde{X}_e W_{q_h}, K_h = \tilde{X}_e W_{k_h} \\
A_h = softmax(\frac{Q_h K_h^T}{\sqrt{d_K}})
\end{aligned}
\tag{1}
$$

where $\tilde{X}_e \in \mathbb{R}^{C \times d_m}$ are the feature embeddings, and W_{q_h} and W_{k_h} are the parameter matrices projecting $\tilde{X}_e$ to query Q_h and key K_h, respectively.

3.3   Graph-Based EEG Classification

As shown in Fig. 1, the final EEG classification is achieved by first adding the head-wise aggregated output from the Chebyshev GNN and EEG feature embeddings from the temporal feature encoder, and average pooling the result along the electrode dimension to yield a final representation of shape $\tilde{X}_g \in \mathbb{R}^{C \times 1}$. A linear layer is then used to perform Healthy vs. PD classification. We use the cross-entropy loss and AdamW optimizer [15] to train our model. Here, we use the Chebyshev GNN in our model, as it has previously been used for EEG analysis [6,15] and is an effective method of integrating an adjacency matrix with EEG feature embeddings by efficiently approximating graph convolutions using Chebyshev polynomials.

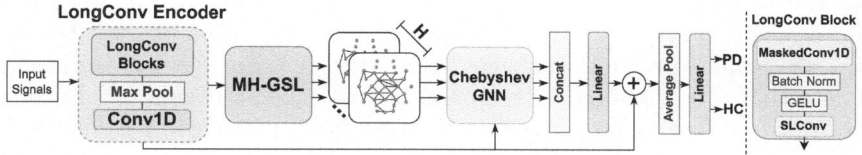

Fig. 1. Overview of the model architecture for PD detection.

3.4   Head-Wise Gradient-Weighted Graph Attention Explainer

In multi-head self-attention networks, the average or maximum of the head-wise attention scores [26] are often used to provide graph explanations, but this could be insufficient as some heads may carry greater contributions for decision-making. Inspired by the work of Rasoulian *et al.* [19], where head-wise gradient-weighted self-attention maps were used to improve the specificity of the attention map, we adapted the core idea for GNN-based EEG analysis. Specifically, we obtain a graph explanation by first weighing the head-wise graph representation A_h with the norm of its gradient based on the class activation. Then, the final adjacency matrix $A \in \mathbb{R}^{C \times C}$ is generated as:

$$A = \frac{1}{H} \sum_{h=1}^{H} ||\frac{\partial Y}{\partial A_h}|| \cdot A_h \qquad (2)$$

where H is the number of attention heads and Y is the target class to generate a graph representation for. Finally, A is thresholded to keep the attention scores within two standard deviations from the mean, and then are normalized to $[0, 1]$.

3.5   Dataset and Preprocessing

We used the UC San Diego Parkinson's disease resting-state EEG (rs-EEG) dataset [20] for our study. The dataset contains the resting-state data of 15 PD patients (63.2 ± 8.2 years, 8 females) and 16 healthy controls (63.5 ± 9.6 years, 9 females). All PD patients had mild to moderate disease severity. Each participant had at least 3 min of resting state data recorded using a 32-channel Biosemi ActiveTwo EEG system (sampling rate = 512 Hz). We minimally preprocessed each subject's EEG by first setting the reference to the mean of the EXG7 and EXG8 mastoid electrodes and band-pass filtered the raw signal to 0.5–80 Hz. The data was then segmented into 2 s of non-overlapping windows, resulting in 90 trials per participant.

3.6   Experimental Setup and Ablation Studies

To assess the classification performance of our proposed framework, we compared it against a variety of DL models and configurations. With CNN methods dominating EEG analysis, as a baseline, we re-implemented the method by Dose *et al.* [7] that showed great success on small datasets. To further validate the benefits of each design component of our method, we performed a series of ablation studies. **First**, to confirm the contribution of the Chebyshev GNN, we compared the full version of our method (CL-Encoder+Freeze) against PD detection only based on the temporal feature encoder (LongConv Encoder). **Second**, to verify whether our multi-head GSL had a positive impact on the network performance, we replaced the learnt graph structure input to the Chebyshev GNN with a static graph based on PCC, and evaluate the classification accuracy against the original design ("Full Model w/o MH-GSL vs. Full Model with MH-GSL", both without CL). **Third**, to quantify the performance gain from the SimCLR framework, we compared the proposed frameworks with and without self-supervised pre-training ("CL-Encoder+Freeze vs. Full Model with MH-GSL"). **Finally**, as some studies demonstrated the benefit of finetuning pre-trained feature encoder, we further tested our proposed method by finetuning the feature encoder weights that were pre-trained using the SimCLR framework, and compared the outcome to freezing the feature encoder weights after SimCLR pre-training ("CL-Encoder+Finetune vs. CL-Encoder+Freeze"). We computed classification accuracy, precision and recall, macro F1-score, and AUC metrics for all experimental setups over 3 random seeds (i.e., model weight initialization).

We trained and evaluated all configurations using a leave-one-out cross-validation, where a single subject was used for testing and the rest for training to avoid data leakage. For each fold, two subjects (one healthy and one PD) were randomly selected from the training data as a validation set. Unlike the more common sample-wise cross-validation in EEG-related DL algorithms, our subject-wise strategy can better assess the generalizability of the proposed framework to unseen subjects. Each model was trained with a batch size of 8 with a MultiStep learning rate (LR) scheduler at an initial LR of 1E−4 and a gamma of 0.1. The MH-GSL model was trained using 2 attention heads and the Chebyshev GNN used a single layer with K = 5 and a dropout rate of 0.2.

4   Results

We present the PD vs. Healthy classification performance of all experiments in Table 1, and with an accuracy of $69.40 \pm 1.59\%$, our proposed method (CL-

Table 1. PD vs. Healthy classification performance for all model configurations.

Method	Accuracy %	AUC	F1-Score	Precision	Recall
LongConv Encoder	64.68 ± 1.85	0.638 ± 0.039	0.643 ± 0.017	0.649 ± 0.020	0.644 ± 0.018
Full Model w/o MH-GSL	66.97 ± 1.29	0.670 ± 0.013	0.663 ± 0.009	0.677 ± 0.021	0.666 ± 0.011
Full Model with MH-GSL	67.73 ± 0.85	**0.715 ± 0.024**	0.672 ± 0.009	0.682 ± 0.009	0.674 ± 0.009
CL-Encoder + Freeze	**69.40 ± 1.59**	0.656 ± 0.036	**0.682 ± 0.016**	**0.716 ± 0.021**	**0.688 ± 0.015**
CL-Encoder + Finetune	66.34 ± 2.68	0.707 ± 0.010	0.658 ± 0.030	0.668 ± 0.026	0.660 ± 0.027
CNN classifier [7]	62.99 ± 4.07	0.640 ± 0.061	0.629 ± 0.040	0.629 ± 0.041	0.629 ± 0.040

Encoder+Freeze) outperformed the CNN baseline [7] (accuracy = 62.99 ± 4.07%) and the other model configurations. For the ablation studies, we confirmed the positive impact of Chebyshev GNN, multi-head graph structure learner, simCLR-based encoder pretraining. Furthermore, between CL-Encoder+Finetune and CL-Encoder+Freeze, further finetuning the feature encoder during full model training decreased all evaluation metrics by 3–5%. In addition, Fig. 2 presents the resulting adjacency matrices averaged for the PD and HC groups for all correctly classified samples based on static PCC-based graphs, mean of head-wise attentions from our MH-GSL, and gradient-weighted mean head-wise attention also from our MH-GSL. The gradient-weighted adjacency matrices show a greater amount of connections towards the inion (back) of the skull compared to their non-weighted counterparts. The PCC graphs show almost exclusively connections between neighboring nodes.

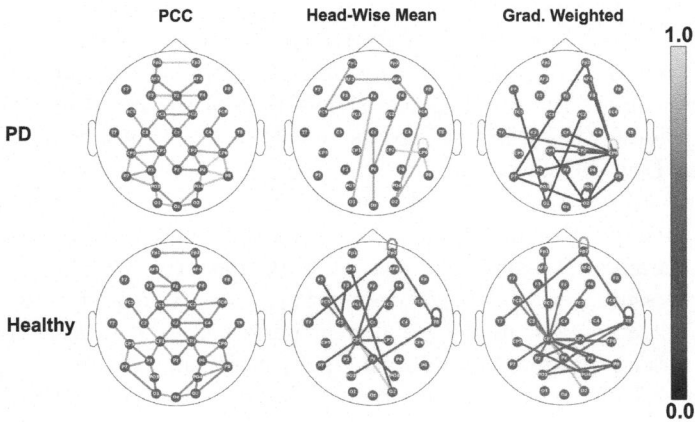

Fig. 2. Group-wise mean adjacency matrices for PD and healthy subjects for static PCC, mean head-wise attention, and gradient-weighted mean head-wise attention.

5   Discussion

Our novel multi-head graph structure learner presents a more dynamic approach that establishes task-driven graphs with improved performance in comparison to static connectivity graphs. This observation agrees with previous studies [23]. So far, despite many attempts to learn graph edge weights using attention mechanisms [2,23], very few extended their formulation to include multiple attention heads despite their great success in vision and language tasks. Different from approaches where attention scores are multiplied with the features in an initial graph [2,13], we directly learn different node features for each adjacency matrix from MH-GSL in parallel, and finally concatenate them for classification. After testing different numbers of attention heads (2, 4 and 8), we found that two heads yielded superior performance for this task. To the best of our knowledge, we are the first to propose a head-wise gradient weighted graph attention explanation to obtain visual interpretation for task-relevant brain connectivity properties. This approach helps further highlight task-relevant graph information. Figure 2 reveals that graphs learnt with our method focus more on global connections across the scalp, and overcome the overemphasis on adjacent connections seen in commonly used stationary graphs. It is also interesting to note that weighing the head-wise adjacency matrices by the norm of their gradients results in a more connected graph structure compared to its unweighted counterpart. Qualitatively, the number of connections seems to greatly increase with gradient-weighing for PD subjects, thus showing a higher connection count to be important for classification. Although an increase in functional connectivity has been shown in PD patients in resting state EEG studies [1], additional analysis of the generated edge explanations is required before drawing neuroscientific conclusions. Nevertheless, the presented technique offers great potential for deriving important connectivity information for the disorder under study. We will further validate the physiological significance of the resulting graph explanation with joint EEG-fMRI studies as the relevant insights could be of more value than PD vs HC classification.

In our experiments, we adopted a subject-wise leave-one-out cross-validation instead of a sample-wise one seen in many reports. The latter approach is often used to accommodate limited subjects in EEG datasets, but can easily cause data leakage issues, resulting in exaggerated accuracy. When adopting this commonly used strategy, our model yields near perfect classification results ($\sim$98% accuracy) potentially due to memorizing subject-specific details instead of task relevant ones. To help address limited data size, we employed contrastive learning to enhance the robustness of our feature encoder, and its benefit is evident in our experiment (1.67% accuracy increase). In comparison to fMRI and task-based EEG, rs-EEG is easier to acquire, but requires more sophisticated feature extraction techniques. Through PD detection, we demonstrated great performance of the proposed DL method and a novel graph explanation technique. We will showcase its adaptability in extended applications in the future.

6   Conclusion

We have developed a novel GNN technique for PD detection from resting state EEG based on dynamic graph structure learning, with a head-wise gradient-weighted graph explainer. In addition, we demonstrated the benefit of contrastive learning in efficient and robust feature extraction from a small cohort. With thorough evaluations and ablation studies, the performance of our proposed method has a great potential to offer clinical insights for PD and extended neurological applications with more accessible EEG sensors.

Acknowledgements. We acknowledge the support of the Natural Sciences and Engineering Research Council of Canada and Y. Xiao is supported by the Fond de la Recherche en Santé du Québec (FRQS-chercheur boursier Junior 1).

Disclosure of Interests. The authors have no competing interests to declare that are relevant to the content of this article.

References

1. Bosch, T.J., Espinoza, A.I., Mancini, M., Horak, F.B., Singh, A.: Functional connectivity in patients with Parkinson's disease and freezing of gait using resting-state EEG and graph theory. Neurorehabil. Neural Repair **36**(10–11), 715–725 (2022)
2. Chang, H., Liu, B., Zong, Y., Lu, C., Wang, X.: EEG-based Parkinson's disease recognition via attention-based sparse graph convolutional neural network. IEEE J. Biomed. Health Inform. **27**(11), 5216–5224 (2023)
3. Chen, T., Kornblith, S., Norouzi, M., Hinton, G.: A simple framework for contrastive learning of visual representations (2020)
4. Covert, I., et al.: Temporal graph convolutional networks for automatic seizure detection, May 2019
5. Demir, A., Koike-Akino, T., Wang, Y., Erdogmus, D.: EEG-GAT: graph attention networks for classification of electroencephalogram (EEG) signals. In: 2022 44th Annual International Conference of the IEEE Engineering in Medicine & Biology Society (EMBC), Glasgow, Scotland, United Kingdom, pp. 30–35. IEEE, July 2022
6. Dissanayake, T., Fernando, T., Denman, S., Sridharan, S., Fookes, C.: Geometric deep learning for subject independent epileptic seizure prediction using scalp EEG signals. IEEE J. Biomed. Health Inform. **26**(2), 527–538 (2022)
7. Dose, H., Møller, J.S., Iversen, H.K., Puthusserypady, S.: An end-to-end deep learning approach to MI-EEG signal classification for BCIs. Expert Syst. Appl. **114**, 532–542 (2018)
8. Gu, A., Goel, K., Ré, C.: Efficiently modeling long sequences with structured state spaces, August 2022. arXiv:2111.00396 [cs]
9. He, J., Cui, J., Zhang, G., Xue, M., Chu, D., Zhao, Y.: Spatial-temporal seizure detection with graph attention network and bi-directional LSTM architecture. Biomed. Signal Process. Control **78**, 103908 (2022)
10. Jin, M., Chen, H., Li, Z., Li, J.: EEG-based emotion recognition using graph convolutional network with learnable electrode relations. In: 2021 43rd Annual International Conference of the IEEE Engineering in Medicine & Biology Society (EMBC), Mexico, pp. 5953–5957. IEEE, November 2021

11. Klepl, D., Wu, M., He, F.: Graph neural network-based EEG classification: a survey, December 2023
12. Lawhern, V.J., Solon, A.J., Waytowich, N.R., Gordon, S.M., Hung, C.P., Lance, B.J.: EEGNet: a compact convolutional network for EEG-based brain-computer interfaces. J. Neural Eng. **15**(5), 056013 (2018)
13. Li, Y., et al.: Dynamical graph neural network with attention mechanism for epilepsy detection using single channel EEG. Med. Biol. Eng. Comput. **62**(1), 307–326 (2024)
14. Li, Y., Cai, T., Zhang, Y., Chen, D., Dey, D.: What makes convolutional models great on long sequence modeling? (2022)
15. Loshchilov, I., Hutter, F.: Decoupled weight decay regularization, January 2019
16. Mohsenvand, M.N., Izadi, M.R., Maes, P.: Contrastive representation learning for electroencephalogram classification. In: Alsentzer, E., McDermott, M.B.A., Falck, F., Sarkar, S.K., Roy, S., Hyland, S.L. (eds.) Proceedings of the Machine Learning for Health NeurIPS Workshop. Proceedings of Machine Learning Research, vol. 136, pp. 238–253. PMLR, 11 December 2020
17. Nerrise, F., Zhao, Q., Poston, K.L., Pohl, K.M., Adeli, E.: An explainable geometric-weighted graph attention network for identifying functional networks associated with gait impairment. In: Greenspan, H., et al. (eds.) MICCAI 2023. LNCS, vol. 14221, pp. 723–733. Springer, Cham (2023). https://doi.org/10.1007/978-3-031-43895-0_68
18. Oord, A.v.d., Li, Y., Vinyals, O.: Representation learning with contrastive predictive coding, January 2019
19. Rasoulian, A., Salari, S., Xiao, Y.: Weakly supervised intracranial hemorrhage segmentation using head-wise gradient-infused self-attention maps from a swin transformer in categorical learning. In: Machine Learning for Biomedical Imaging (MLCN 2022), vol. 2, pp. 338–360, August 2023
20. Rockhill, A.P., Jackson, N., George, J., Aron, A., Swann, N.C.: UC San Diego resting state EEG data from patients with Parkinson's disease (2021)
21. Song, T., Zheng, W., Song, P., Cui, Z.: EEG emotion recognition using dynamical graph convolutional neural networks. IEEE Trans. Affect. Comput. **11**(3), 532–541 (2020)
22. Sun, M., Cui, W., Yu, S., Han, H., Hu, B., Li, Y.: A dual-branch dynamic graph convolution based adaptive transformer feature fusion network for EEG emotion recognition. IEEE Trans. Affect. Comput. **13**(4), 2218–2228 (2022)
23. Tang, S., et al.: Modeling multivariate biosignals with graph neural networks and structured state space models. In: Mortazavi, B.J., Sarker, T., Beam, A., Ho, J.C. (eds.) Proceedings of the Conference on Health, Inference, and Learning. Proceedings of Machine Learning Research, 22 June–24 June 2023, vol. 209, pp. 50–71. PMLR
24. Tolosa, E., Garrido, A., Scholz, S.W., Poewe, W.: Challenges in the diagnosis of Parkinson's disease. Lancet Neurol. **20**(5), 385–397 (2021)
25. Vaswani, A., et al.: Attention is all you need (2023)
26. Veličković, P., Cucurull, G., Casanova, A., Romero, A., Liò, P., Bengio, Y.: Graph attention networks (2018)
27. Vetter, J., Macke, J.H., Gao, R.: Generating realistic neurophysiological time series with denoising diffusion probabilistic models, August 2023

ProxiMO: Proximal Multi-operator Networks for Quantitative Susceptibility Mapping

Shmuel Orenstein[1], Zhenghan Fang[1,4], Hyeong-Geol Shin[2,3], Peter van Zijl[1,2,3], Xu Li[2,3], and Jeremias Sulam[1,4(✉)]

[1] Department of Biomedical Engineering, Johns Hopkins University, Baltimore, MD 21218, USA
{sorenst3,zfang23,jsulam1}@jhu.edu
[2] F.M. Kirby Research Center for Functional Brain Imaging, Kennedy Krieger Institute, Baltimore, MD 21205, USA
[3] Department of Radiology and Radiological Sciences, Johns Hopkins University, Baltimore, MD 21205, USA
[4] Johns Hopkins Kavli Neuroscience Discovery Institute, Baltimore, MD 21218, USA

Abstract. Quantitative Susceptibility Mapping (QSM) is a technique that derives tissue magnetic susceptibility distributions from phase measurements obtained through Magnetic Resonance (MR) imaging. This involves solving an ill-posed dipole inversion problem, however, and thus time-consuming and cumbersome data acquisition from several distinct head orientations becomes necessary to obtain an accurate solution. Most recent (supervised) deep learning methods for single-phase QSM require training data obtained via multiple orientations. In this work, we present an alternative unsupervised learning approach that can efficiently train on single-orientation measurement data alone, named ProxiMO (Proximal Multi-Operator), combining Learned Proximal Convolutional Neural Networks (LP-CNN) with multi-operator imaging (MOI). This integration enables LP-CNN training for QSM on single-phase data without ground truth reconstructions. We further introduce a semi-supervised variant, which further boosts the reconstruction performance, compared to the traditional supervised fashions. Extensive experiments on multi-center datasets illustrate the advantage of unsupervised training and the superiority of the proposed approach for QSM reconstruction. Code is available at https://github.com/shmuelor/ProxiMO

Keywords: Quantitative Susceptibility Mapping · Deep Learning · Inverse Problems · Unsupervised Learning

1 Introduction

Quantitative Susceptibility Mapping (QSM) is a technique in Magnetic Resonance Imaging (MRI) that focuses on the quantification of tissue magnetic

D. R. Bathula et al. (Eds.): MLCN 2024, LNCS 15266, pp. 13–23, 2025.
https://doi.org/10.1007/978-3-031-78761-4_2

susceptibility, derived from the phase information of gradient echo imaging [1,2]. QSM holds significant clinical potential due to its ability to offer valuable insights into tissue composition and microstructure [2,3], including characteristics like myelin content in white matter and iron deposition in gray matter. Pathological alterations in these sources of tissue contrast are intricately linked to a range of neurodegenerative disorders, including but not limited to Multiple Sclerosis [4–6] and Alzheimer's Disease [7–11].

QSM processing typically involves two central processing steps [12]. The first regards phase preprocessing, including phase unwrapping and background field removal, yielding a tissue frequency image as a result. The second, more challenging step is the dipole inversion, i.e. reconstructing the susceptibility map from the tissue frequency information. Because of the singularity present in the dipole kernel and the constrained quantity of phase measurements within the field of view, the process of inverting the dipole kernel is an ill-posed inverse problem [12]. Different methods exist to address this inversion problem. The standard approach involves over-sampling, wherein multiple phase measurements collected from diverse head orientations are used to partially resolve the ill-posedness. This technique, known as calculation of susceptibility through multiple orientation sampling (COSMOS), is commonly employed as gold standard for in vivo QSM [13]. Despite its high image quality, COSMOS is often prohibitive since phase images from three or more head orientations are needed, resulting in long scan times, high costs, and discomfort for patients.

With the surge of deep learning in medical imaging, recent QSM algorithms based on neural networks have provided approaches to approximate the dipole inversion process, even when supplied with only a single phase measurement [14]. Different methods, based on different architectures and loss functions, have been proposed [15–18]. We focus on the Learned Proximal Convolutional Neural Network (LP-CNN) approach, an iterative proximal gradient-like approach [18] to carry out QSM reconstruction. LP-CNN provides state-of-the-art results when compared to both traditional techniques and deep learning-based methods, as well as better generalization via the integration of the forward operator. All of these methods, however, employ COSMOS reconstruction from multiple phase measurements as ground-truth training data in a fully-supervised training manner. This strong dependence on multiple-angle orientations acquisitions severely restricts the quantity of data that can be used for training, as most datasets with COSMOS have only about a dozen patients. On the other hand, large datasets like the Human Connectome Project [19] or BIOCARD [20,21] have hundreds or thousands of scans, but only at one or a couple of orientations, which is insufficient to provide COSMOS reconstructions. Alternative, unsupervised learning methods for QSM have not yet been thoroughly explored. The recent AdaIN-QSM work [22] introduced an unsupervised QSM deep learning method, with competitive performance compared to leading classical methods. Their approach reconstructs QSM with variable resolutions through adaptive instance normalization [23], applicable across resolutions. We will comment on further relations on this work later in Sect. 2.

In this work, we introduce ProxiMO, an algorithm inspired in the framework of LP-CNN but that can be trained when ground-truth (or COSMOS) data is limited, or even inaccessible altogether. This is achieved by integrating LP-CNN with a recent learning formulation termed multi-operator imaging (MOI) [24]. MOI leverages data from a group of diverse forward operators to learn data-driven solutions to inverse problems in an unsupervised manner, learning from incomplete measurements. By combining MOI with LP-CNN, ProxiMO provides an unsupervised learning approach for QSM that can leverage large-scale datasets (which do not include COSMOS data) with just a single orientation of phase measurement. Furthermore, we extend MOI to semi-supervised scenarios where COSMOS data is partially available on a subset of training subjects. We show that even in that case, the combination of MOI and LP-CNN outperforms the traditional supervised training setting of LP-CNN. We train and deploy ProxiMO on three in-vivo human datasets acquired from different institutions, and demonstrate its ability to improve QSM reconstruction.

2   Methods

In QSM, the physics model between tissue magnetic susceptibility and tissue frequency changes can be represented by $y = \Phi x + \nu$, where $y \in \mathcal{Y} \subset \mathbb{R}^n$ represents the tissue frequency derived from phase measurement, $x \in \mathcal{X} \subset \mathbb{R}^n$ denotes the underlying susceptibility map, and $\mathcal{Y}$ and $\mathcal{X}$ correspond to the spaces of phase measurements and susceptibility maps, respectively. The map $\Phi : \mathcal{X} \to \mathcal{Y}$ serves as the forward operator and ν accounts for measurement noise and potential model mismatches. The forward operator can be expressed as $\Phi = F^{-1}DF$, with F representing the discrete Fourier transform and D being a diagonal matrix known as the dipole kernel [25,26]. The goal of dipole inversion is to reconstruct susceptibility map x from the phase y. Given that the dipole kernel D contains zero elements on its diagonal, the operator Φ is singular [25]. Thus, one must resort to regularizing this ill-posed problem, as in the following formulation:

$$\min_x \frac{1}{2}\|y - \Phi x\|_2^2 + \psi(x), \tag{1}$$

where $\psi : \mathbb{R}^n \to \mathbb{R}$ serves as a regularizer. In this way, solving the QSM phase-to-susceptibility dipole inversion boils down to minimizing the problem above, resulting in an estimate of the susceptibility map, which we will denote as $\hat{x}$.

2.1   Unsupervised QSM via ProxiMO

LP-CNN [18] provides an appealing solution to address the phase-to-susceptibility dipole inversion through a data-driven approach, taking inspiration from the Proximal Gradient Descent algorithm [27] with a data-dependent operator – in lieu of a proximal – implemented via a 3D Wide ResNet [28]. Denote the estimation produced by LP-CNN as $\hat{x} = g_\theta(y, \Phi)$, where θ represents the parameters of the model. For training, the optimization formulation involves minimizing the ℓ_2 error in reconstruction over the training samples $\{y_j, x_j^c\}$, where x^c is the COSMOS reconstruction of x. This formulation can be expressed as follows:

$$L_{\text{supervised}} = \frac{1}{N} \sum_{j=1}^{N} \|x_j^c - g_\theta(y_j, \Phi_j)\|_2^2, \qquad (2)$$

where N is the number of training samples. Naturally, the above loss function (and those of most other deep learning-based approaches) requires the COSMOS reconstruction x^c for each subject, which significantly limits the amount of data that can be used for model training. In this work, we aim to adapt LP-CNN to an unsupervised framework via multi-operator imaging (MOI) from [24].

MOI [24] is an approach designed to address inverse problems as those addressed in this work when only incomplete measurement data are available for training. In general, unsupervised learning using a fixed incomplete measurement process is impossible, as there is no information about the underlying data distribution in the null space of the measurement operator. MOI overcomes this limitation by leveraging *multiple* measurement operators, $\Phi_1, ..., \Phi_G$, which provide complementary information about the data distribution along different subspaces. To effectively train on such data, MOI uses an unsupervised loss function that promotes coherence across all these measurement projections, i.e.:

$$L_{\text{unsupervised}} = \frac{1}{N} \sum_{j=1}^{N} \left[\|y_j - \Phi_j \hat{x}_j\|_2^2 + \|\hat{x}_j - g_\theta(\Phi' \hat{x}_j, \Phi')\|_2^2 \right], \qquad (3)$$

where $\hat{x}_j = g_\theta(y_j, \Phi_j)$ and Φ' is a forward operator with a randomly selected angle[1]. This loss function comprises two components. The initial part ensures the consistency between reconstruction and observed measurements, thus promoting $y_j \approx \Phi_j \hat{x}_j$. On the other hand, the second part enforces consistency among operators, that is, $\hat{x}_j \approx g_\theta(\Phi' \hat{x}_j, \Phi')$. By using such a MOI loss, we can train a model for QSM reconstruction without needing access to any COSMOS data. This eliminates a fundamental constraint on the applicability of LP-CNN and other supervised learning methods for QSM. It is worthwhile comparing this formulation to that in recent work of AdaIN-QSM [22]. Their work only implemented the first part of our MOI loss (Eq. (3)), i.e., the measurement consistency part, with a different motivation, but does not include the second term that enforces cross-operator consistency. Additionally, their method used other losses that are not present in ProxiMO, such as total variation and gradient difference loss, which can be combined with MOI training to further improve performance in the future.

2.2   ProxiMO Allows for Semi-supervised QSM

Integrating MOI into LP-CNN proves beneficial not only in scenarios where COSMOS data is completely unavailable but also when COSMOS is limited and

---

[1] I.e., we take a direction at random by initially sampling a 3D i.i.d. Gaussian vector and then projecting it onto the unit sphere.

accessible for just a small number of subjects. In such cases, we propose employing a convex combination of the supervised loss in Eq. (2) and unsupervised MOI loss in Eq. (3), i.e.: $L_{\text{semi-supervised}} = \gamma L_{\text{supervised}} + (1 - \gamma) L_{\text{unsupervised}}$, where $\gamma \in [0, 1]$ controls a weight between both losses. Naturally, $L_{\text{supervised}}$ can only be computed for the data associated with COSMOS reconstructions; otherwise, it will be zero. $L_{\text{unsupervised}}$ can be computed for samples only with their phase measurements, y_j at a single orientation. By employing this loss function, we can achieve enhanced utilization of an available COSMOS dataset. The requirement for cross-operator consistency in terms of MOI leads to improved performance, surpassing what is achievable with a conventional LP-CNN framework. It is important to emphasize that MOI was originally introduced as an unsupervised loss [24]. To the best of our knowledge, incorporating MOI into a semi-supervised framework represents a novel application of this technique.

3   Experiments and Results

In this section, we demonstrate the application of ProxiMO to provide QSM reconstruction trained on a large dataset where no ground-truth (COSMOS) is available, and then showcase its performance when trained on a combination of data with and without COSMOS reconstructions. In all cases, ProxiMO provides state-of-the-art single-orientation QSM estimation.

3.1   Data Acquisition and Implementation Details

All the experiments reported in this paper were conducted using three distinct datasets. The first dataset corresponds to the one utilized in the LP-CNN paper [18]. A total of 36 MR phase measurements were acquired at 7T with a Philips Achieva (32 channel head coil) scanner, using three slightly different 3D gradient echo (GRE) sequences with a resolution of $1 \times 1 \times 1\,\text{mm}^3$ on 8 healthy subjects (5 orientations for 4 subjects with $\text{TR} = 28$ ms, $\text{TE1}/\delta\text{TE} = 5/5$ ms, 5 echoes, $\text{FOV} = 224 \times 224 \times 126\,\text{mm}^3$, 4 orientations for 3 subjects with $\text{TR} = 45$ ms, $\text{TE1}/\delta\text{TE} = 2/2$ ms, 9 echoes, $\text{FOV} = 224 \times 224 \times 110\,\text{mm}^3$, 4 orientations for 1 subject with $\text{TR} = 45$ ms, $\text{TE1}/\delta\text{TE} = 2/2$ ms, 16 echoes, FOV $= 224 \times 224 \times 110\,\text{mm}^3$) at F.M. Kirby Research Center for Functional Brain Imaging, Kennedy Krieger Institute. The study was IRB-approved (informed consents were obtained from all participants). For all subjects, GRE magnitude data at each orientation were first coregistered to a natural supine position using FSL FLIRT [29]. The dipole kernel for each head orientation was calculated from acquisition and orientation parameters, and the tissue frequency maps were transformed to a natural supine position via the co-registration transform matrices. Best-path-based phase unwrapping [30], brain masking with FSL BET [31], and V-SHARP [32] for removing the background field, and echo averaging for echoes with TEs between 10 ms and 30 ms, were applied for all cases, as stated in [18]. The COSMOS reconstructions were then generated using all 4 or 5 orientations for each subject, as available. Following [15], we included data augmentation via 36 simulated tissue frequency maps at random head orientations (within $\pm 45°$ of the z-axis) by using the COSMOS calculated susceptibility source.

The second dataset [33] encompasses 12 subjects, each also accompanied by a COSMOS reconstruction. A total of 65 MR phase measurements were acquired at 3T with a resolution of $1 \times 1 \times 1\,\text{mm}^3$ on 12 healthy subjects (6 orientations for 11 subjects, 11 orientations for 1 subject, all of them with $TR = 38\,\text{ms}$, $TE1/\delta TE = 7.7/5$ ms, 6 echoes, $FOV = 256 \times 224 \times 160\,\text{mm}^3$). Lastly, we utilized the BIOCARD dataset [20,21], which consists of 523 QSM scans, each with only one orientation, rendering it unfeasible to perform COSMOS reconstruction for each subject.

Our code was implemented in Pytorch and the models were trained on a single NVIDIA GeForce RTX 2080 Ti GPU with a mini-batch of size 2. To avoid the requirement of a large amount of GPU memory, we partition our data into $64 \times 64 \times 64$ 3D patches for training, as was done in LP-CNN training [18]. To prevent performance degradation caused by the loss of crucial high-frequency information, zero-padding is incorporated to the forward operator.

3.2   ProxiMO in an Unsupervised Manner

We first demonstrate ProxiMO unsupervised, employing the SGD optimizer with an initial learning rate of 10^{-3}. Given the absence of COSMOS data, we compare with TKD [34] and the application of the pseudo-inverse naive solution, which is the regularized solution to the least squares problem, given by $\hat{x} = (\Phi^H \Phi + \lambda I)^{-1} \Phi^H y$, with $\lambda = 0.01$ and H is the Hermitian adjoint[2]. Unfortunately, the recent unsupervised method, AdaIN-QSM [22], has not provided an implementation code. However, we implemented and reproduced their algorithm according to their paper and compared our results with theirs.

Table 1. Quantitative performance metrics for unsupervised QSM reconstruction.

Methods	Validation Set		Test Set	
	PSNR (dB)	SSIM	PSNR(dB)	SSIM
TKD	34.92 ± 0.413	0.9719 ± 0.0019	34.93 ± 0.562	0.9818 ± 0.0021
Pseudo-Inverse	37.53 ± 0.838	0.9809 ± 0.0025	37.42 ± 0.880	0.9860 ± 0.0025
AdaIN-QSM [22]	38.16 ± 0.360	0.9853 ± 0.0012	37.05 ± 0.957	0.9870 ± 0.0028
ProxiMO	38.66 ± 0.672	0.9869 ± 0.0021	38.22 ± 0.812	0.9907 ± 0.0018

Since ProxiMO can employ data at only single orientations, we trained our model using 499 BIOCARD QSM scans, including six out of the eight subjects

---

[2] The parameter λ was fine-tuned to optimize the performance of the pseudo-inverse method.

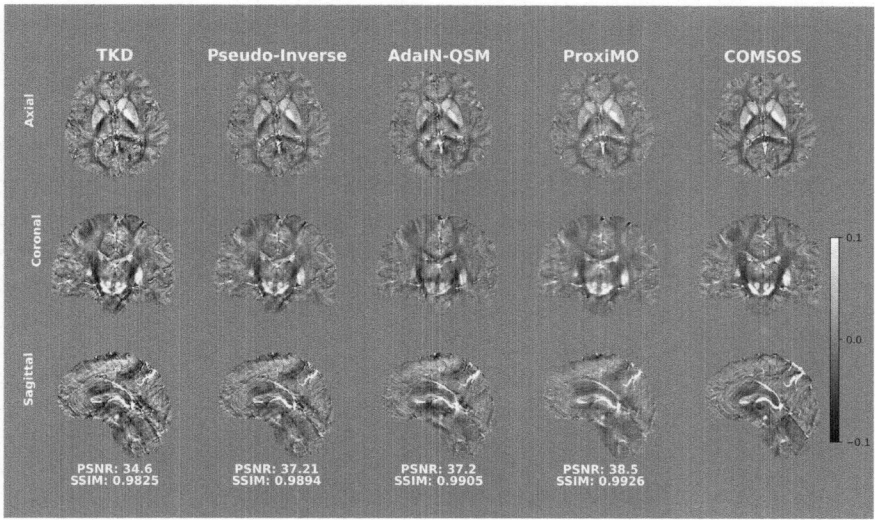

Fig. 1. A comparison of susceptibility maps reconstructed using various methods, including ProxiMO, in an unsupervised manner.

from the first dataset (still without the COSMOS reconstructions). Two subjects were held back for validation. The evaluation of our model's performance was carried out on the second dataset. During training, we employed phase measurements from only two orientations per subject, selected randomly if multiple orientations were available. This approach aimed to simulate a realistic scenario where the acquisition of a limited number of orientations precluded the execution of COSMOS. The numerical results are shown in Table 1, where visual results are shown in Fig. 1. Indeed, ProxiMO outperforms TKD and the Pseudo-Inverse, yielding higher PSNR and SSIM.

3.3 ProxiMO in a Semi-supervised Manner

We now showcase the capabilities of ProxiMO in a semi-supervised manner. In this case, we trained the model with Adam optimizer and an initial learning rate of 10^{-4}. Two subjects from the first dataset were designated as validation subjects, whereas the entire second dataset was designated as the test set. The remaining 6 subjects from the first dataset were denoted as T and used for training. Let n denote the number of subjects in the training set for which COSMOS has been obtained. For each $n \in [5]$, six subsets of subjects $T' \subseteq T$ of size $|T'| = n$ were randomly sampled as the subjects with COSMOS. The remaining $6 - n$ subjects were treated as subjects in the training set where COSMOS is not available, and phase measurements from only two orientations were randomly selected from each subject to mimic the realistic scenario where only a small number of orientations were acquired and COSMOS could not be performed. These phase measurements, combined with all the phase measurements in T', were used as the unsupervised training data for ProxiMO. For each T' at each n, both an LP-CNN model and a ProxiMO model were trained. The training set for the LP-CNN model comprised only the n subjects in T' (i.e., subjects with COSMOS). In contrast, the ProxiMO model was trained on not only the n subjects with COSMOS but also the remaining $6 - n$ subjects without COS-MOS via the semi-supervised ProxiMO loss. Similar procedures were performed for $n = 6$, whereas only one subset T' was selected, as $\binom{6}{6} = 1$. The γ value in our combined loss $L_{\text{semi-supervised}}$ was increased linearly from 0.5 to 0.8 as n goes from 1 to 6.

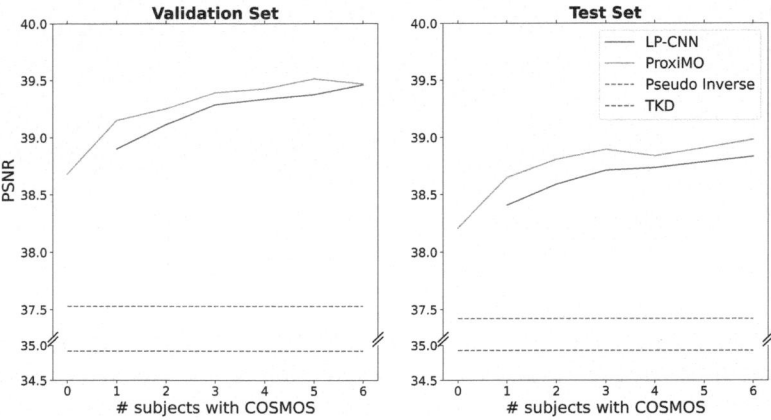

Fig. 2. Average outcome of the semi-supervised experiments with each point repre-senting the average of 6 experiments. Note that at 0 subjects, LP-CNN reconstruction is not available.

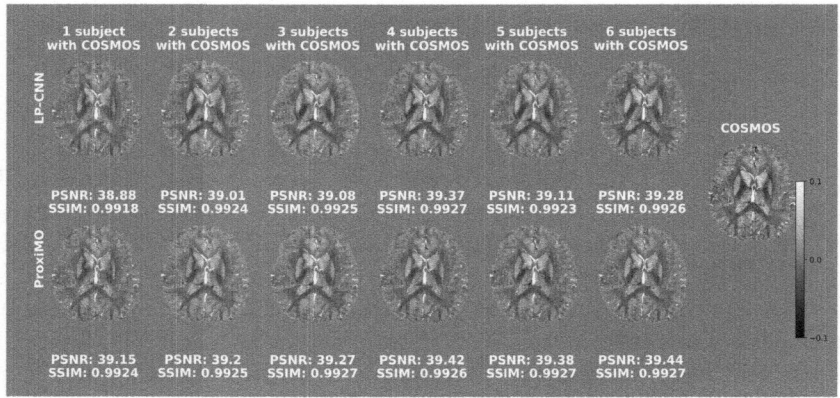

Fig. 3. A comparison between LP-CNN and semi-supervised ProxiMO.

Overall, we conducted 31 pairs of experiments. For each experiment, we performed model selection on the validation set and tested the model on the test set. The results of all experiments are depicted in Fig. 2, whereas visual results are shown in Fig. 3. The advantage of the PSNR and SSIM of ProxiMO over LP-CNN is clear. As ProxiMO uses the same dataset for training, this improvement shows that the ProxiMO semi-supervised setting is a better utilization of a given dataset that includes COSMOS, as the ProxiMO loss function looks not only on the quality of a specific reconstruction (as in Eq. (2)) but also on its cross-operator consistency (as in $L_{\text{semi-supervised}}$).

4   Conclusion

In this study, we introduced ProxiMO, a Proximal Multi-Operator method offering an unsupervised solution for the challenging dipole inversion problem in QSM. Our method improves the scalability of supervised QSM techniques like LP-CNN by removing the requirement for ground-truth data during training. Additionally, we showcased that ProxiMO can not only handle scenarios where COSMOS data is absent but can also improve the results of supervised methods trained with COSMOS data present.

Acknowledgements. This research has been supported by NIH, USA Grant P41EB031771, as well as by the Toffler Charitable Trust, USA, and by the Distinguished Graduate Student Fellows program of the KAVLI Neuroscience Discovery Institute, USA.

References

1. De Rochefort, L., Brown, R., Prince, M.R., Yi, W.: Quantitative MR susceptibility mapping using piece-wise constant regularized inversion of the magnetic field. Magn. Reson. Med. **60**(4), 1003–1009 (2008)
2. Wang, Y., Liu, T.: Quantitative susceptibility mapping (QSM): decoding MRI data for a tissue magnetic biomarker. Magn. Reson. Med. **73**(1), 82–101 (2015)
3. Liu, C., Li, W., Tong, K.A., Yeom, K.W., Kuzminski, S.: Susceptibility-weighted imaging and quantitative susceptibility mapping in the brain. J. Magn. Reson. Imaging **42**(1), 23–41 (2015)
4. Chen, W., et al.: Quantitative susceptibility mapping of multiple sclerosis lesions at various ages. Radiology **271**(1), 183–192 (2014)
5. Langkammer, C., et al.: Quantitative susceptibility mapping in multiple sclerosis. Radiology **267**(2), 551–559 (2013)
6. Li, X., et al.: Magnetic susceptibility contrast variations in multiple sclerosis lesions. J. Magn. Reson. Imaging **43**(2), 463–473 (2016)
7. Acosta-Cabronero, J., Williams, G.B., Cardenas-Blanco, A., Arnold, R.J., Lupson, V., Nestor, P.J.: In vivo quantitative susceptibility mapping (QSM) in Alzheimer's disease. PLoS ONE **8**(11), e81093 (2013)
8. Ayton, S., et al.: Cerebral quantitative susceptibility mapping predicts amyloid-β-related cognitive decline. Brain **140**(8), 2112–2119 (2017)

9. Van Bergen, J.M.G., et al.: Colocalization of cerebral iron with amyloid beta in mild cognitive impairment. Sci. Rep. **6**(1), 35514 (2016)
10. Cogswell, P.M., Fan, A.P.: Multimodal comparisons of QSM and PET in neurodegeneration and aging. Neuroimage **273**, 120068 (2023)
11. Ravanfar, P., et al.: Systematic review: quantitative susceptibility mapping (QSM) of brain iron profile in neurodegenerative diseases. Front. Neurosci. **15**, 41 (2021)
12. Deistung, A., Schweser, F., Reichenbach, J.R.: Overview of quantitative susceptibility mapping. NMR Biomed. **30**(4), e3569 (2017)
13. Liu, T., et al.: Calculation of susceptibility through multiple orientation sampling (COSMOS): a method for conditioning the inverse problem from measured magnetic field map to susceptibility source image in MRI. Magn. Reson. Med. **61**(1), 196–204 (2009)
14. Jung, W., Bollmann, S., Lee, J.: Overview of quantitative susceptibility mapping using deep learning: current status, challenges and opportunities. NMR Biomed. **35**(4), e4292 (2022)
15. Yoon, J., et al.: Quantitative susceptibility mapping using deep neural network: QSMNet. Neuroimage **179**, 199–206 (2018)
16. Bollmann, S., et al.: DeepQSM-using deep learning to solve the dipole inversion for quantitative susceptibility mapping. Neuroimage **195**, 373–383 (2019)
17. Feng, R., et al.: MODL-QSM: model-based deep learning for quantitative susceptibility mapping. Neuroimage **240**, 118376 (2021)
18. Lai, K.-W., Aggarwal, M., van Zijl, P., Li, X., Sulam, J.: Learned proximal networks for quantitative susceptibility mapping. In: Martel, A.L., et al. (eds.) MICCAI 2020. LNCS, vol. 12262, pp. 125–135. Springer, Cham (2020). https://doi.org/10.1007/978-3-030-59713-9_13
19. Van Essen, D.C., et al.: The WU-Minn human connectome project: an overview. Neuroimage **80**, 62–79 (2013)
20. Albert, M., et al.: Cognitive changes preceding clinical symptom onset of mild cognitive impairment and relationship to ApoE genotype. Curr. Alzheimer Res. **11**(8), 773–784 (2014)
21. Chen, L., et al.: Quantitative susceptibility mapping of brain iron and β-amyloid in MRI and pet relating to cognitive performance in cognitively normal older adults. Radiology **298**(2), 353–362 (2021)
22. Oh, G., Bae, H., Ahn, H.-S., Park, S.-H., Moon, W.-J., Ye, J.C.: Unsupervised resolution-agnostic quantitative susceptibility mapping using adaptive instance normalization. Med. Image Anal. **79**, 102477 (2022)
23. Huang, X., Belongie, S.: Arbitrary style transfer in real-time with adaptive instance normalization. In: IEEE International Conference on Computer Vision, pp. 1501–1510 (2017)
24. Tachella, J., Chen, D., Davies, M.: Unsupervised learning from incomplete measurements for inverse problems. Adv. Neural. Inf. Process. Syst. **35**, 4983–4995 (2022)
25. Liu, C., Wei, H., Gong, N.-J., Cronin, M., Dibb, R., Decker, K.: Quantitative susceptibility mapping: contrast mechanisms and clinical applications. Tomography **1**(1), 3–17 (2015)
26. Ruetten, P.P.R., Gillard, J.H., Graves, M.J.: Introduction to quantitative susceptibility mapping and susceptibility weighted imaging. Br. J. Radiol. **92**(1101), 20181016 (2019)
27. Boyd, S., Parikh, N.: Proximal algorithms. Found. Trends Optim. **1**(3), 123–231 (2013)

28. Zagoruyko, S., Komodakis, N.: Wide residual networks. *arXiv preprint* arXiv:1605.07146 (2016)
29. Jenkinson, M., Smith, S.: A global optimisation method for robust affine registration of brain images. Med. Image Anal. **5**(2), 143–156 (2001)
30. Abdul-Rahman, H., et al.: Fast three-dimensional phase-unwrapping algorithm based on sorting by reliability following a non-continuous path. In: Optical Measurement Systems for Industrial Inspection IV, vol. 5856, pp. 32–40. SPIE (2005)
31. Smith, S.M.: Fast robust automated brain extraction. Hum. Brain Mapp. **17**(3), 143–155 (2002)
32. Özbay, P.S., et al.: A comprehensive numerical analysis of background phase correction with V-sharp. NMR Biomed. **30**(4), e3550 (2017)
33. Shin, H.-G., et al.: chi-separation using multi-orientation data invivo and exvivo brains: visualization of histology up to the resolution of 350 μm. In: Joint Annual Meeting ISMRM-ESMRMB & ISMRT 31st Annual Meeting, London, UK (2022)
34. Shmueli, K., de Zwart, J.A., van Gelderen, P., Li, T.-Q., Dodd, S.J., Duyn, J.H.: Magnetic susceptibility mapping of brain tissue in vivo using MRI phase data. Magn. Reson. Med. **62**(6), 1510–1522 (2009)

Brain-Cognition Fingerprinting via Graph-GCCA with Contrastive Learning

Yixin Wang[1], Wei Peng[2], Yu Zhang[3], Ehsan Adeli[1], Qingyu Zhao[4(✉)],
and Kilian M. Pohl[2]

[1] Department of Bioengineering, Stanford University, Stanford, CA, USA
[2] Department of Psychiatry and Behavioral Sciences, Stanford University, Stanford, CA, USA
[3] Department of Bioengineering, Lehigh University, Bethlehem, PA, USA
[4] Department of Radiology, Weill Cornell Medicine, New York, USA
qiz4006@med.cornell.edu

Abstract. Many longitudinal neuroimaging studies aim to improve the understanding of brain aging and diseases by studying the dynamic interactions between brain function and cognition. Doing so requires accurate encoding of their multidimensional relationship while accounting for individual variability over time. For this purpose, we propose an unsupervised learning model (called <u>Co</u>ntrastive Learning-based <u>Gra</u>ph Generalized <u>Ca</u>nonical Correlation Analysis (CoGraCa)) that encodes their relationship via Graph Attention Networks and generalized Canonical Correlational Analysis. To create brain-cognition fingerprints reflecting unique neural and cognitive phenotype of each person, the model also relies on individualized and multimodal contrastive learning. We apply CoGraCa to longitudinal dataset of healthy individuals consisting of resting-state functional MRI and cognitive measures acquired at multiple visits for each participant. The generated fingerprints effectively capture significant individual differences and outperform current single-modal and CCA-based multimodal models in identifying sex and age. More importantly, our encoding provides interpretable interactions between those two modalities.

1 Introduction

Longitudinal neuroimaging studies often repeatedly acquire functional MRI and neurocognitive performance measures of study participants to explore the connection between brain function and cognition and their development over time [11,13,16]. The investigations, however, are often hindered by the lack of computational tools linking such multi-modal, repeated measures. Despite the advances of machine learning in neuroimaging studies, existing models [6,32,33] are often designed to predict univariate outcome variables, which cannot characterize shared and dissociated neural bases underlying multiple cognitive domains (e.g.,

---

Supplementary Information The online version contains supplementary material available at https://doi.org/10.1007/978-3-031-78761-4_3.

working memory, motor functions). Moreover, cross-sectional methods often fail to disentangle the consistent brain functional connectivity of a single subject across multiple visits, from the variations that exist between individuals [3,5].

One potential solution to relating multi-dimensional functional and cognitive measures is Canonical Correlation Analysis (CCA) [26], which has been successfully applied in a number of cross-sectional studies to identify reproducible brain-cognition mapping. Traditional CCA methods [26] primarily capture linear associations between modalities, which are not suitable for modeling the complex spatial characteristics inherent in brain connectivity. An approach that has been highly successful in inferring neural activity patterns in functional MRI are Graph Neural Networks (GNNs) [10,14,20,21], which have been coupled with the CCA framework to relate two augmented views derived from fMRI signals for spurious factor mitigation [21]. Thus, GNN-based CCA might provide a strong foundation for learning brain-cognition mapping, but it remains unclear how such mappings preserve inter-subject variability and intra-subject consistency.

In this work, we propose **Co**ntrastive Learning-based **Gra**ph Generalized **Ca**nonical Correlation Analysis (CoGraCa), aimed at encoding the correlation between brain functional connectivity and cognitive measurements at individual-level while characterizing brain functional differences to create personalized brain-cognition fingerprints that reflect the unique neural and cognitive landscapes of each person. We utilize a Graph Attention Network (GAT) to encode brain functional connectivity derived from resting-state functional MRI (rs-fMRI). The GAT is coupled with a generalized CCA (GCCA) [2], which jointly encodes brain function and cognitive scores so that the resulting brain functional networks are aligned with the cognitive data. To explicitly account for the inter-subject variability and intra-subject consistency, we further design two contrastive learning strategies: i) An individualized contrastive learning approach that regulates the graph embeddings both within and between subjects in the latent space, ensuring that the unique connectivity patterns of each subject are preserved while differentiating between subjects. ii) A longitudinal multimodal contrastive learning that encourages the cross-modal alignment of brain connectivity and cognitive measures across different visits within each subject, maintaining the dynamic evolution of individualized brain-cognition correlation.

Our proposed CoCraCa is cross-validated on a dataset comprising 57 participants, totaling 93 visits containing both fMRI and cognitive measures. The generated "brain-cognition" fingerprints demonstrate significant individual differentiation. Validated through downstream sex and age classification task using these fingerprints, CoGraCa achieves higher accuracy scores in comparison to other state-of-the-art single-modal and CCA-based multimodal methods, underscoring its effectiveness in integrating brain connectivity with cognitive data for precise individual characterization. Importantly, CoGraCa enables interpretable correlations between modalities, identifying sex- and age-related functional connectivity and cognitive measures that align with established neuroscience research.

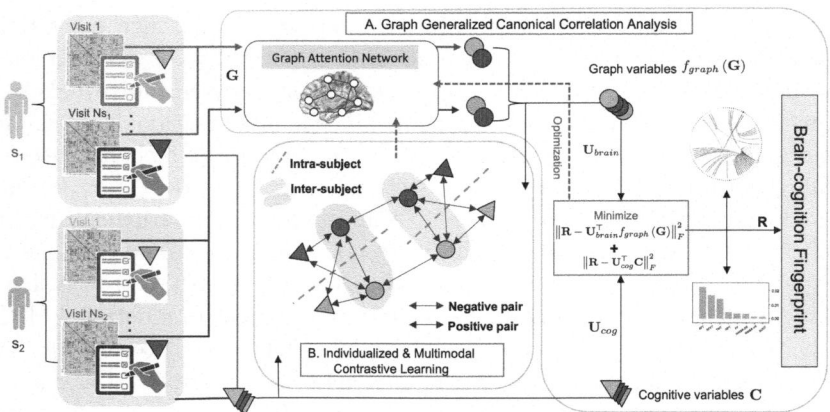

Fig. 1. Overview of our model. (A) Brain functional connectivity of each subject is encoded using graph attention networks (GAT) into graph embedding. The learned embedding (graph variables) and cognitive measures (cognitive variables) are mapped to a shared "brain-cognition" space via generalized canonical correlation analysis. The GAT weights are updated by optimizing the maximum correlation between two modalities. (B) An individualized contrastive learning differentiates inter-subjects brain connectivity and a multimodal contrastive learning to align brain connectivity and cognitive measures across multiple visits within subjects, capturing intra-subject and cross-modal dynamics.

2   Method

Let S be the number of subjects and N be the number of visits across all subjects, with N_s representing the number of visits for subject s. Each visit i contains a set of cognitive measures $\mathcal{C}_i$ and a connectivity matrix encoding the Pearson correlation between the fMRI signal of brain Regions of Interests (ROIs). The connectivity matrix is represented as a graph $\mathcal{G}_i = (\mathbf{A}_{\mathcal{G}_i}, \mathbf{D}_{\mathcal{G}_i})$ consisting of V nodes (representing the ROIs). $\mathbf{A}_{\mathcal{G}_i} \in \mathbb{R}^{V \times V}$ is the adjacent matrix consisting only of positive correlations (anticorrelations are set to 0) [15,30]. $\mathbf{D}_{\mathcal{G}_i} \in \mathbb{R}^{V \times D}$ is the attribute matrix, represented by each ROI's "connection profile" of length D, as defined in [4]. Based on the set of pairs $\{\mathbf{G}, \mathbf{C}\} = \{(\mathcal{G}_1, \mathcal{C}_1), (\mathcal{G}_2, \mathcal{C}_2), \cdots, (\mathcal{G}_n, \mathcal{C}_n)\}$, the goal of our approach is to learn a brain-cognition representation (or "Brain-Cognition fingerprints") $\mathbf{R} = \{\mathcal{R}_1, \mathcal{R}_2, \cdots, \mathcal{R}_n\}$. We determine the optimal $\mathbf{R}$ by regularizing Graph GCCA (Fig. 1 A) by individualized and multimodal contrastive learning (Fig. 1 B). We now describe these components in further detail.

Graph Generalized Canonical Correlation Analysis. For each input connectivity graph $\mathcal{G}_i$, we adopt Graph Attention Layers (GAT) [28] to learn its encoding into a node embedding $\boldsymbol{h}_{\mathcal{G}_i} \in \mathbb{R}^{V \times r}$, where the embedding $\boldsymbol{h}_p \in \mathbb{R}^{1 \times r}$ of node p is updated by aggregating the features of its 1-hop neighborhood nodes $\mathcal{N}_p$ through self-attention mechanism. Specifically, the attention between p and its neighboring node q at layer k is calculated by: $a_{pq}^k \left(\boldsymbol{h}_p^k, \boldsymbol{h}_q^k \right) =$

Softmax $\left(\sigma \left(m^T \left[\mathbf{W} \boldsymbol{h}_p^k \| \mathbf{W} \boldsymbol{h}_q^k \right] \right) \right)$, where $\|$ denotes a concatenation operation, σ is a non-linear activation function ReLU, m is a trainable single-layer feed-forward neural network, and W is a trainable weight matrix. The node representation at layer $k+1$ will be further obtained by: $\boldsymbol{h}_p^{k+1} = \sigma \left(\sum_{q \in \mathcal{N}_p} a_{pq}^k \mathbf{W} \boldsymbol{h}_q^k \right)$. The node embeddings $\{\boldsymbol{h}_p\}_{p \in V}$ of each $\mathcal{G}_i$ from the last layer are further applied to a global mean pooling operation to obtain a set of graph variables $\boldsymbol{h}_{\mathcal{G}_i}$. We aim to learn this encoding function $\mathcal{H} := f_{graph}(\mathbf{G})$ through maximizing its correlation with cognitive measures. Specifically, the representations $\mathcal{H} := \{\boldsymbol{h}_{\mathcal{G}_i}\}_{\mathcal{G}_i \in \mathbf{G}}$ learned from the complex brain connectivity, along with the cognitive measures, are treated as two sets of canonical variables that will be projected into a shared "brain-cognition" space to obtain $\mathbf{R}$ through GCCA [2,8]. Note, the cognitive measures are not subjected to any encoder to ensure its direct guidance to fMRI generation and integration.

The unsupervised CCA optimization is expressed as maximizing the sum of correlations between $\mathbf{R}$ and each modality, defined by the loss function: $\mathcal{L}_{corr} = \left\| \mathbf{R} - \mathbf{U}_{brain}^\top f_{graph}(\mathbf{G}) \right\|_F^2 + \left\| \mathbf{R} - \mathbf{U}_{cog}^\top \mathbf{C} \right\|_F^2$, s.t. $\mathbf{R}\mathbf{R}^\top = \mathbf{I}$. $\mathbf{U}_{brain}$ and $\mathbf{U}_{cog}$ are linear transformation matrix (canonical loadings) of the two variables from brain connectivity and cognition measures. Note, $\mathbf{R}$ is the shared representation in the "brain-cognition" space, which is obtained by solving an eigenvalue problem. Following [2], for the sets of brain graphs $\mathbf{G}$, we define the covariance matrix as $\mathbf{Cov} = f_{graph}(\mathbf{G}) f_{graph}(\mathbf{G})^\top$ and obtain the positive semi-definite matrix $\mathbf{P}_{brain} = f_{graph}(\mathbf{G})^\top \mathbf{Cov}^{-1} f_{graph}(\mathbf{G})$. Similarly, we obtain $\mathbf{P}_{cog}$ from $\mathbf{C}$ and we stack them as $\mathbf{M} = \mathbf{P}_{brain} + \mathbf{P}_{cog}$. Then, the eigenvectors of $\mathbf{M}$ will be constructed into $\mathbf{R}$ which maximally and linearly correlates non-linear transformations of brain functional connectivity and cognition measures. This optimization problem is solved by estimating the gradient of the objective on samples that are mapped through f_{graph} and using back-propagation to update weights within f_{graph}.

Individualized Contrastive Learning. To capture the inherent individual variability present in brain functional connectivity, we design an individualized contrastive learning strategy where pairs of brain connectivity are constructed from all N visits across S subjects. Pairs from the same subject s are considered to be similar and thus are labeled as positive pairs (see Fig. 1B). Conversely, pairs from different individuals are likely to be dissimilar and are labeled negative accordingly. Specifically, given the sets of graph embeddings $\mathcal{H} := \{\boldsymbol{h}_{\mathcal{G}_i}\}_{\mathcal{G}_i \in \mathbf{G}}$, we define a subject indicator $\mathbb{K}_{\boldsymbol{y}_{ij}=1}$, where $\boldsymbol{y}_{ij} = 1$ denotes a pair of graph embeddings $\boldsymbol{h}_{\mathcal{G}_i}$ and $\boldsymbol{h}_{\mathcal{G}_j}$ are from the same subject, otherwise, $\boldsymbol{y}_{ij} = 0$. The individualized contrastive loss can then be achieved by

$\mathcal{L}_{\text{ind}} = -\frac{1}{N} \sum_{i \in N} \sum_{j \in N} \mathbb{K}_{\boldsymbol{y}_{ij}=1} \log \frac{\exp\left(\text{sim}\left(\boldsymbol{h}_{\mathcal{G}_i}, \boldsymbol{h}_{\mathcal{G}_j}\right)/\tau\right)}{\sum_{k \in N, k \neq i} \exp\left(\text{sim}\left(\boldsymbol{h}_{\mathcal{G}_i}, \boldsymbol{h}_{\mathcal{G}_k}\right)/\tau\right)}$, where $\text{sim}(\cdot)$ denotes the cosine similarity and $\tau > 0$ is a temperature parameter that controls the separation of subjects. By doing so, graph embeddings from the same subject s are pulled closer together than embeddings from different subjects.

Multimodal Contrastive Learning. Meanwhile, we have to ensure the above individual-level separation keeps the longitudinal differences within each subject. To achieve this goal, we apply a multimodal contrastive learning, inspired by CLIP [22], yet specifically tailored to within-subject pairs for our multimodal features, i.e., brain connectivity and cognitive measures, to capture this intra-subject and cross-modal dynamics. For subject $s \in S$ with the number of visits $N_s > 1$, given the paired brain graph $\mathcal{G}_i^s$ and cognitive measures $\mathcal{C}_i^s$ across N_s visits, we aim to maximize their similarity from the same visit and minimize the similarity from different visits within each subject. This multimodal contrastive learning is achieved by: $\mathcal{L}_{\text{mul}} =$

$$-\frac{1}{S} \sum_{s \in S} \sum_{i \in N_s} \log \frac{\exp\left(\text{sim}\left(h_{\mathcal{G}_i^s}, \mathcal{C}_i^s\right)/\tau\right)}{\sum_{k \in N_s, k \neq i} \exp\left(\text{sim}\left(h_{\mathcal{G}_i^s}, \mathcal{C}_k^s\right)/\tau\right)}.$$

By accounting for both the brain connectivity variability between subjects and variability in correlation with cognitive measures within each subject, the model can derive a more individualized representation that integrates the longitudinal variations in brain connectivity specific to cognitive measures. The final objective function is defined as $\mathcal{L}_{\text{total}} := \mathcal{L}_{\text{corr}} + \lambda_1 \mathcal{L}_{\text{ind}} + \lambda_2 \mathcal{L}_{\text{mul}}$, where λ_1 and λ_2 are trade-off parameters to balance the two contrastive learning procedures.

3   Experimental Results

Dataset. Our study utilizes the SRI dataset (PIs: Pfefferbaum and Sullivan) consisting of rs-fMRI (3T GE Discovery MR750 scanner, 8-channel head coil, echo time=30ms, dwell-time=0.388ms, TR=2200ms, 2.5mm isotropic after upsampling) of 417 subjects (822 visits). Of them, 195 participants (275 visits) completed cognitive testing at the same visit as the rs-fMRI was acquired. The cognitive measurements are summarized in 16 domain-specific scores. Of the 195 subjects, 57 subjects (89 visits, age: 58.53±10.57 years) are normal controls with 21 females (33 visits) and 36 males (56 visits). Of each of the 89 rs-fMRI, the pipeline by [7] extracts connectivity matrix across 111 ROIs. Each entry in that matrix is the Pearson correlation between the rs-fMRI signals of two ROIs.

Implementation Details. We implement the proposed model CoGraCa using PyTorch with the Adam optimizer and a learning rate of 0.001. Our graph encoder is composed of two GAT layers, with hidden units=32. The dimension of the node embedding r is set as 16 and the number of canonical variants (i.e., dimension of $\mathbf{R}$) are set as 16. τ is set as 0.9 and λ_1 and λ_2 are set as 1.5 and 0.5, respectively. The model is trained for 1000 epochs using five-fold cross-validation with folds defined by subjects to ensure that all visits from a single subject are assigned to the same fold. For each test fold, the model is trained on the remaining data to optimize the model's parameters and obtain canonical loadings $\mathbf{U}_{brain}, \mathbf{U}_{cog}$. After training is completed, the "Brain-Cognition" representation $\mathbf{R}$ is generated for each sample from the test fold. Codes will be made available at https://github.com/Wangyixinxin/BrainCog.

3.1   Individual Variability of "Brain-Cognition" Fingerprints

We assess whether our derived representations could capture individual-specific features more effectively than other CCA-based methods.

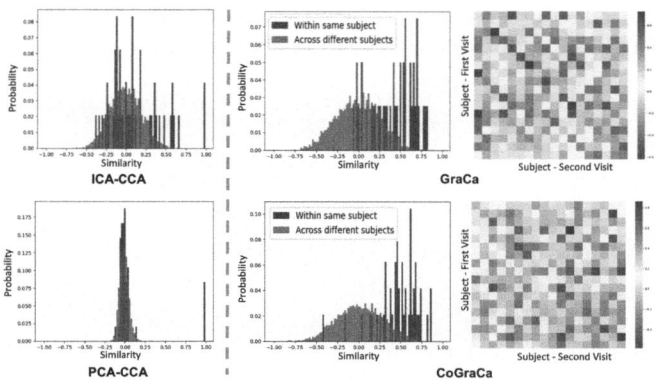

Fig. 2. Histograms show the similarity in the representations between visits within subject (intra-subject, red) vs. across subjects (inter-subject, blue). Of all methods, CoGraCa model produced the most individualized representation, i.e., the intra-subject similarity is relatively high compared to the inter-subject similarity (Color figure online).

Experimental Setup. For comparison, we repeat the five fold cross-validation by applying Principal Component Analysis (PCA) and Independent Component Analysis (ICA) to the connectivity matrices before performing CCA analysis. PCA reduces the connectivity matrices to 544 independent components (which accounted for 95% of the data variance) and ICA to 20 independent components (chosen based on the best performance compared with 10,15,20,25 components). Separately for PCA and ICA, the components are fed into CCA in conjunction with cognitive measures to obtain a representation, labelled as PCA-CCA and ICA-CCA in our comparison. Finally, we run CoGraCa omitting contrastive learning (referred to as GraCa). For each model, we compute the similarity between each pair of representations both within subjects across different visits and between subjects across their visits using Pearson correlation.

Results. The histograms in Fig. 2 for ICA-CCA and PCA-CCA show that the distributions of within-subject similarity (red) and between-subject similarity (blue) largely overlap, indicating a lack of individual distinctiveness in the representations generated by these models. The two distributions are much better separated by the representations generated by GraCa and CoGraCa, with individuals being significantly distinct from each other (Mann-Whitney U test, CoGraCa: p-value < 0.0001, GraCa: p-value < 0.0001). CoGraCa is also associated with a larger Wasserstein distance (0.45), i.e., larger distribution separation, compared with GraCa (0.39). We also measure the similarity between visits from the same and different subjects for 18 subjects with two visits in Fig. 2 (See GraCa and CoGraCa). The correlation matrices reveal that CoGraCa produces highly differentiable individualized representations, where each participant exhibits a high correlation with their own across visits (as reflected in the diagonal of correlation matrix) and low correlations from visits from other

Table 1. Balanced accuracies on sex and age prediction tasks. Results were averaged across 5 folds and run 10 times with random seeds. The best results are shown in **bold**.

	Sex Classification	Age Classification
Functional MRI-only		
ICA	56.05±2.3	69.38±2.24
Vanilla GCN [12]	54.73±2.43	58.75±2.57
BrainGNN [14]	61.25±2.12	63.57±1.09
GAT-LI [10]	58.22±1.59	62.73±1.99
Cognition-only	73.45±1.24	71.81±2.01
multimodal		
PCA-CCA	72.93±1.56	71.31±2.01
ICA-CCA	65.42± 2.27	63.31±2.14
SDGCCA [18]	69.11±1.94	65.7±2.57
GraCa	74.14±1.46	70.30±1.52
CoGraCa	**74.90±1.78**	**73.26±1.21**

cohorts, leading to individualized "brain-cognition" fingerprints. GraCa yields highly similar representations within individual participants but also displays a higher degree of similarity to other subjects than CoGraCa.

3.2 Validating Fingerprints with Downstream Tasks

We conduct a quantitative analysis to determine if the integrated "brain-cognition" representations from CoGraCa enhance the accuracy of identifying specific individual characteristics compared to GraCa and other CCA-based multimodal methods that aim to correlate brain function and cognition. Based on the trained model for each of the 5 test folds, the representations for both the training and testing sets are generated for downstream tasks without any additional fine-tuning of the model.

The downstream tasks focus on predicting age (older vs younger) and sex as brain function and cognition often differentiate between these cohorts [1,27]. Given the relatively small sample size of our data set, the task of age prediction is confined to younger ($\leq$60 years, 47 visits, 38.3% are females) versus older (¿60 years, 42 visits, 35.7% are females) as in [24]. The sex ratio is similar between those two cohorts according to Chi-square test (p=0.97). For identifying sex, males (age: 59.21±10.01 years) and females (age: 57.38±11.68 years) have similar age (t-test, p=0.45). The classification model is a multi-layer perceptron (MLP) containing two fully-connected layers of dimension 64 and 32 with ELU and a dropout rate of 0.5. Due to the limitation of the small data size, we repeat the cross-validation of the MLP 10 times using different seed points for initialization. For each cross-validation, we record the balanced accuracy (BACC).

Baseline. We compare our method against single modality-based (fMRI-only and cognition-only) and multimodal CCA-based approaches. With respect to

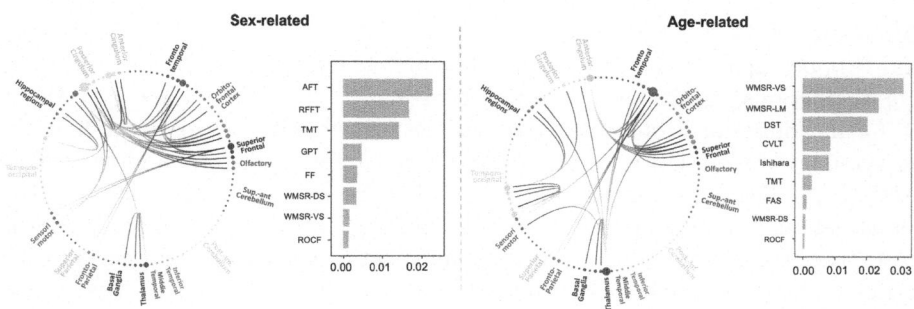

Fig. 3. Identified functional connectivity and rank-ordered positive loadings of cognitive variables of task-related CCA component, in line with [9,17,19,23,29,31]. Functional regions and cognitive variables were detailed in Supplemental Fig. S4.

fMRI-only, we derive representations via ICA (20 components) and perform classification with MLP. Supervised methods include vanilla GCN [12], BrainGNN [14], and GAT-LI [10]. Multimodal methods were PCA-CCA and ICA-CCA. We also employ the state-of-the-art supervised SDGCCA approach [18]. Each method is trained and tested using the same experimental setup as for GraCa and CoGraCa.

Results. Table 1 lists the average and standard deviation accuracy scores across the 10 runs. Solely relying on correlation matrices (i.e., fMRI) results in relatively low scores regardless of representation due to the low signal-to-noise ratio of that modality. The accuracy scores of the multimodal baseline methods (PCA-CCA, ICA-CCA, and SDGCCA) are higher but also lower than only relying on the cognitive measures, which suggests that they are not able to properly integrate the multimodal data. Our method achieves that goal and has the highest BACC for sex and age. All its accuracy scores are higher than GraCa, which aligns with our expectation that multimodal contrastive learning effectively maintains the longitudinal intra-individual distinctions.

3.3 Functional Connectivity and Cognition Interpretation

A significant achievement of our approach is the extraction of cognitive-related functional connectivities and identifying the most relevant features from both modalities linked to sex or age distinctions. Using SHapley Additive exPlanations (SHAP) [25], we identify key features in "brain-cognition" representations that drive predictions. These features correspond to specific CCA components, leading us to extract canonical loadings U_{brain} and U_{cog} that highlight the significance of graph and cognitive variables.

By combining graph variables with the attention matrices obtained from GAT that indicate the learned connectivity from CoGraCa, we derive the functional connectivity pattern identifying sex and age (Fig. 3). Color-coding of brain functional regions is defined according to [7] and their contribution is encoded by the node size. With respect to sex, our method identifies dense connectivity within

Orbito-frontal Cortex (OFC), Frontotemporal (FT), Posterior Cingulum (PCC), and Hippocampal (HPC) regions, which is in line with the literature [29,31]. For age distinctions, significant connections involve Temporo-occipital (TO), HPC, OFC, and Superior Frontal (SF) regions, resonating with neuroscience findings on age-related neural alterations on Parahippocampal, Occipital and Prefrontal areas [9,23]. See Supplemental Fig. S1 and Fig. S2 for additional insights on the robustness of connectivities across folds and Fig. S3 for connectivity patterns from other CCA components. These brain functional patterns interacting with cognitive measures are in line with the literature: Alternate Finger Tapping Test (AFT) is important for identifying sex [19] and, for age, the Wechsler Memory Scale-Revised Test (WMSR) [17], which assesses visual/logical memory. Interestingly, WMSR is correlated correctly with functional regions related to memory (e.g. TO and HPC) revealing that CoCraCa provides a meaningful integration between brain function and cognition.

4 Conclusion

In this work, we introduced a novel unsupervised approach, CoGraCa, to accurately encode brain function coupled with cognition as captured by longitudinal rs-fMRI and cognitive testing. CoGraCa generates "brain-cognition" fingerprints capturing the unique neural and cognitive landscapes of individuals across time by coupling Graph GCCA with individualized and multimodal contrastive learning. We measure the accuracy of CoGraCa by using the encoding to identify the sex and age in individuals. Our multimodal encoding has a higher balanced accuracy than several state-of-the-art representations. More importantly, CoGraCa allows us to identify the brain-cognition relationship important for these tasks.

Acknowledgments. The work was partly funded by the National Institute of Health (DA057567, AA05965, AA017347, AA010723, MH129694, MH130956, AG080425, AA028840), the DGIST Joint Research Project, the 2024 Stanford HAI Hoffman-Yee Grant, the Stanford HAI-Google Cloud Credits Award, BBRF Young Investigator Grant and the Lehigh University FIG (FIGAWD35) and CORE (001250) grants.

References

1. Bachmann, D., et al.: Age-, sex-, and pathology-related variability in brain structure and cognition. Trans. psychiatry **13**(1) (2023)
2. Benton, A., Khayrallah, H., Gujral, B., Reisinger, D.A., Zhang, S., Arora, R.: Deep generalized canonical correlation analysis. In: Proceedings of the 4th Workshop on Representation Learning for NLP, Florence, Italy, August 2, 2019, pp. 1–6. Association for Computational Linguistics (2019)
3. Bijsterbosch, J., Harrison, S., Duff, E., Alfaro-Almagro, F., Woolrich, M., Smith, S.: Investigations into within-and between-subject resting-state amplitude variations. Neuroimage **159**, 57–69 (2017)
4. Cui, H., et al.: BrainGB: a benchmark for brain network analysis with graph neural networks. IEEE Trans. Med. Imaging **42**(2), 493–506 (2022)

5. Finn, E.S., Todd Constable, R.: Individual variation in functional brain connectivity: implications for personalized approaches to psychiatric disease. Dialogues Clin. Neurosci. **18**(3), 277–287 (2016)
6. Gao, M., et al.: Multimodal brain connectome-based prediction of suicide risk in people with late-life depression. Nature Mental Health **1**(2), 100–113 (2023)
7. Honnorat, N., et al.: Alcohol use disorder and its comorbidity with HIV infection disrupts anterior cingulate cortex functional connectivity. Biol. Psychiatry: Cogn. Neurosci. Neuroimaging **7**(11), 1127–1136 (2022)
8. Horst, P.: Generalized canonical correlations and their applications to experimental data. J. Clin. Psychol. **17**, 331–47 (1961)
9. Hsieh, S., Yang, M.H., Yao, Z.F.: Age differences in the functional organization of the prefrontal cortex: analyses of competing hypotheses. Cereb. Cortex **33**(7), 4040–4055 (2023)
10. Hu, J., Cao, L., Li, T., Dong, S., Li, P.: GAT-LI: a graph attention network based learning and interpreting method for functional brain network classification. BMC Bioinformatics **22**(1), 1–20 (2021)
11. Ji, J., et al.: Mapping brain-behavior space relationships along the psychosis spectrum. Elife 10 (2021)
12. Kipf, T.N., Welling, M.: Semi-supervised classification with graph convolutional networks. In: 5th International Conference on Learning Representations, ICLR 2017, Toulon, France, April 24-26, 2017, Conference Track Proceedings (2017)
13. Lee, K., et al.: Human brain state dynamics reflect individual neuro-phenotypes. bioRxiv (2023)
14. Li, X., et al.: BrainGNN: interpretable brain graph neural network for fMRI analysis. Med. Image Anal. **74**, 102233 (2021)
15. Li, Y., Wei, Q., Adeli, E., Pohl, K.M., Zhao, Q.: Joint graph convolution for analyzing brain structural and functional connectome. In: Medical Image Computing and Computer Assisted Intervention - MICCAI 2022 - 25th International Conference, Singapore, September 18-22, 2022, Proceedings, Part I. Lecture Notes in Computer Science, vol. 13431, pp. 231–240. Springer (2022)
16. Luo, L., et al.: Patterns of brain dynamic functional connectivity are linked with attention-deficit/hyperactivity disorder-related behavioral and cognitive dimensions. Psychol. Med. pp. 1–12 (2023)
17. Margolis, R.B., Scialfa, C.T.: Age differences in Wechsler memory scale performance. J. Clin. Psychol. **40**(6), 1442–1449 (1984)
18. Moon, S., Hwang, J., Lee, H.: SDGCCA: supervised deep generalized canonical correlation analysis for multi-omics integration. J. Comput. Biol. **29**(8), 892–907 (2022)
19. Morrison, M.W., Gregory, R.J., Paul, J.J.: Reliability of the finger tapping test and a note on sex differences. Percept. Mot. Skills **48**(1), 139–142 (1979)
20. Nerrise, F., Zhao, Q., Poston, K.L., Pohl, K.M., Adeli, E.: An explainable geometric-weighted graph attention network for identifying functional networks associated with gait impairment. In: Medical Image Computing and Computer Assisted Intervention - MICCAI 2023 - 26th International Conference, Vancouver, BC, Canada, Proceedings, Part II. Lecture Notes in Computer Science, vol. 14221, pp. 723–733. Springer (2023)
21. Peng, L., Wang, N., Xu, J., Zhu, X., Li, X.: GATE: graph CCA for temporal self-supervised learning for label-efficient fMRI analysis. IEEE Trans. Med. Imaging **42**(2), 391–402 (2022)

22. Radford, A., et al.: Learning transferable visual models from natural language supervision. In: Proceedings of the 38th International Conference on Machine Learning, 18-24 July 2021, Virtual Event. Proceedings of Machine Learning Research, vol. 139, pp. 8748–8763 (2021)
23. Ramanoël, S., York, E., Le Petit, M., Lagrené, K., Habas, C., Arleo, A.: Age-related differences in functional and structural connectivity in the spatial navigation brain network. Front. Neural Circuits **13**, 69 (2019)
24. Statsenko, Y., et al.: Predicting age from behavioral test performance for screening early onset of cognitive decline. Front. Aging Neurosci. **13**, 661514 (2021)
25. Sundararajan, M., Najmi, A.: The many Shapley values for model explanation. In: Proceedings of the 37th International Conference on Machine Learning, ICML 2020, 13-18 July 2020, Virtual Event. Proceedings of Machine Learning Research, vol. 119, pp. 9269–9278 (2020)
26. Thompson, B.: Canonical correlation analysis. Read. Underst. MORE Multivar. Stat. 285–316 (2000)
27. Tomasi, D., Volkow, N.D.: Measures of brain connectivity and cognition by sex in us children. JAMA Netw. Open **6**(2), e230157–e230157 (2023)
28. Velickovic, P., et al.: Graph attention networks. stat. **1050**(20), 10–48550 (2017)
29. Weis, S., Hodgetts, S., Hausmann, M.: Sex differences and menstrual cycle effects in cognitive and sensory resting state networks. Brain Cogn. **131**, 66–73 (2019)
30. Weissenbacher, A., Kasess, C., Gerstl, F., Lanzenberger, R., Moser, E., Windischberger, C.: Correlations and anticorrelations in resting-state functional connectivity MRI: a quantitative comparison of preprocessing strategies. Neuroimage **47**(4), 1408–1416 (2009)
31. Zhang, C., Dougherty, C.C., Baum, S.A., White, T., Michael, A.M.: Functional connectivity predicts gender: evidence for gender differences in resting brain connectivity. Hum. Brain Mapp. **39**(4), 1765–1776 (2018)
32. Zhang, Y., et al.: Identification of psychiatric disorder subtypes from functional connectivity patterns in resting-state electroencephalography. Nature biomedical engineering **5**(4), 309–323 (2021)
33. Zhu, X., Du, X., Kerich, M., Lohoff, F.W., Momenan, R.: Random forest based classification of alcohol dependence patients and healthy controls using resting state MRI. Neurosci. Lett. **676**, 27–33 (2018)

HyperBrain: Anomaly Detection for Temporal Hypergraph Brain Networks

Sadaf Sadeghian[1(✉)], Xiaoxiao Li[2,3], and Margo Seltzer[1]

[1] Department of Computer Science, University of British Columbia, Vancouver, Canada
{sadafsdn,mseltzer}@cs.ubc.ca
[2] Department of Electrical and Computer Engineering, University of British Columbia, Vancouver, Canada
[3] Vector Institute, Toronto, Canada

Abstract. Identifying unusual brain activity is a crucial task in neuroscience research, as it aids in the early detection of brain disorders. It is common to represent brain networks as graphs, and researchers have developed various graph-based machine learning methods for analyzing them. However, the majority of existing graph learning tools for the brain face a combination of the following three key limitations. First, they focus only on pairwise correlations between regions of the brain, limiting their ability to capture synchronized activity among larger groups of regions. Second, they model the brain network as a static network, overlooking the temporal changes in the brain. Third, most are designed only for classifying brain networks as healthy or disordered, lacking the ability to identify abnormal brain activity patterns linked to biomarkers associated with disorders. To address these issues, we present HyperBrain, an unsupervised anomaly detection framework for temporal hypergraph brain networks. HyperBrain models fMRI time series data as temporal hypergraphs capturing dynamic higher-order interactions. It then uses a novel customized temporal walk (BrainWalk) and neural encodings to detect abnormal co-activations among brain regions. We evaluate the performance of HyperBrain in both synthetic and real-world settings for Autism Spectrum Disorder and Attention Deficit Hyperactivity Disorder (ADHD). HyperBrain outperforms all other baselines on detecting abnormal co-activations in brain networks. Furthermore, results obtained from HyperBrain are consistent with clinical research on these brain disorders. Our findings suggest that learning temporal and higher-order connections in the brain provides a promising approach to uncover intricate connectivity patterns in brain networks, offering improved diagnosis. Our code is available at: https://github.com/ubc-systopia/HyperBrain.

1 Introduction

The brain is an intricate system, and functional magnetic resonance imaging (fMRI) is a widely-used neuroimaging technique for studying brain activity.

Supplementary Information The online version contains supplementary material available at https://doi.org/10.1007/978-3-031-78761-4_4.

Researchers often interpret fMRI data as a simple graph with nodes representing regions of interest (ROI) and edges indicating functional connectivity through pairwise correlations of Blood-Oxygen-Level Dependent (BOLD) time series signals. Recent advances in machine learning methods for analyzing graph-structured data have led to the development of effective approaches for studying human brain networks, particularly in tasks such as disease detection [13,19]. These approaches classify brain states as healthy or indicative of a disorder. However, a crucial step in understanding symptoms and improving early detection of neurobiological disorders is identifying abnormal patterns in the brain. To address this gap, some studies [1,19] focus on brain network classification. They employ statistical tests or significant scores to pinpoint the most crucial regions or pairwise brain connections linked to identifying disorders. However, these approaches have drawbacks, such as depending heavily on classification accuracy, requiring a well-balanced dataset, which is rare in neuroimaging field and overlooking more complex patterns and structural features.

Anomaly detection in the human brain is a promising solution for abnormal brain pattern discovery [7,14,31], but it is a challenging task due to the lack of ground truth labeled anomalies and the need for a powerful brain modeling and analysis approach capable of capturing different patterns in the brain. Many existing anomaly detection methods are not designed for brain networks [3,31,33]. These methods can typically analyze only a single brain in isolation, making them inappropriate for fMRI due to the noisy data and the need of analyzing a group of brains to properly comprehend the disorder and capture dependable group-level biomarkers. Some other methods designed for brain anomaly detection use non-learnable and fixed rules as anomalies, which are not powerful and generalizable enough for the complex nature of brain activity and capturing patterns outside of the defined rules [7]. Furthermore, most of these modeling approaches are limited in two different ways.

A limitation in previous fMRI based brain-modeling studies is that they often assumed that brain networks are static. However, recent research demonstrates dynamic changes in the brain [6], both in task-based fMRI [15] and resting-state fMRI [16], revealing the dynamic nature and biologically meaningful evolution of brain activity. Consequently, researchers have developed methods to track brain activity over time, including extracting dynamic functional connectivity or a temporal graph from fMRI time series [12,16] and using recurrent networks on the fMRI time series [30]. By analyzing dynamic information, they can improve detection accuracy of brain disorder. However, leveraging dynamic information and temporal brain patterns in anomaly detection is under-explored.

Another modeling limitation in many previous brain analysis methods is their predominant focus on simple graphs [4,14,19], ignoring the group activation of ROIs. Clinical research indicates that cognitive mechanisms in the brain involve interactions among multiple co-activated brain regions, not just among pairs [18]. Although others have improved brain classification accuracy using *hypergraphs* to capture the complex relationships among ROIs by introducing hyperedges that connect multiple nodes simultaneously [20,27,32], they often neglect temporal

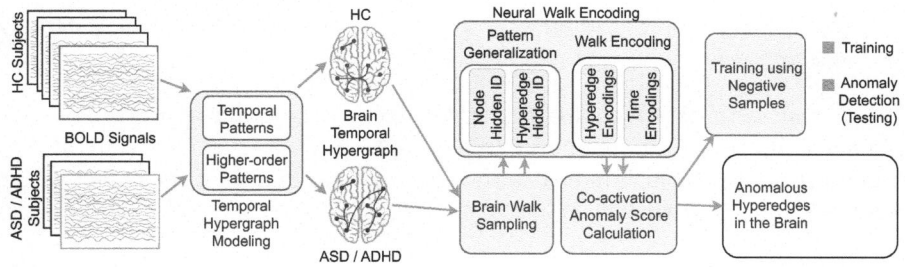

Fig. 1. Schematic of HyperBrain. HyperBrain consists of four stages: (1) Modelling both temporal and higher-order interactions among brain regions [Sect. 2.1], (2) Extracting temporal, higher-order patterns of brain activity [Sect. 2.2], (3) Neural encoding to merge information from the sequence of hyperedges and their timestamps in the extracted brain patterns [Sect. 2.3], and (4) Calculating anomaly scores for each brain co-activation [Sect. 2.3]. In training HyperBrain, we only rely on healthy control data to detect anomalous co-activations in the brain [Sect. 2.4].

patterns [27,32,36] or limit the sizes of higher-order interactions, e.g., considering only interactions among three regions [36]. Moreover, the crucial task of identifying abnormal patterns remains unaddressed [20,27,36].

We present HyperBrain, a specialized framework for detecting abnormal co-activations in brain networks. HyperBrain represents fMRI data as temporal hypergraphs, effectively capturing dynamic higher-order interactions in the brain. It then uses a novel temporal walk customized for brain networks, BrainWalk, to extract higher-order temporal motifs. Then, it learns the structural and temporal properties of brain networks through neural encodings for higher-order walks. Finally, HyperBrain uses these encodings to calculate an anomaly score for each co-activation. By leveraging a training approach on diverse healthy brain networks, HyperBrain enhances robust learning and mitigates noise, enabling it to identify anomalous hyperedges in the brains of individuals with disorders. Remarkably, HyperBrain only relies on the neuroimaging data from healthy control group for training, eliminating the need for a balanced dataset of healthy and disordered subjects.

Our experiments highlight HyperBrain's outstanding performance in detecting abnormal brain co-activation associated with Attention Deficit Hyperactivity Disorder (ADHD) and Autism, outperforming all other baseline methods. Furthermore, our real-world experiments show HyperBrain's ability to detect abnormal brain activity. Figure 1 illustrates the components of HyperBrain.

2 The Proposed Framework

Definition 1 (Temporal Hypergraph). *A temporal hypergraph is defined as* $\mathcal{G} = (\mathcal{V}, \mathcal{E})$, *where* $\mathcal{V}$ *denotes the set of nodes, and* $\mathcal{E}$ *represents hyperedges occurring in the hypergraph over time. Specifically,* $\mathcal{E}$ *is defined as the set*

$\mathcal{E} = \{(e_1, t_1), (e_2, t_2), \dots\}$, where $e_i \in 2^{\mathcal{V}}$ represents a hyperedge, and t_i denotes the timestamp when e_i occurs.

Our task is to detect anomalous hyperedges in the brain network. For each hyperedge in the brain temporal hypergraph $(e_k, t_k) \in \mathcal{E}$, we compute an anomaly score $\varphi(e)$ which indicates the level of abnormality of the co-activation represented by e_k at time t_k.

2.1   Modeling fMRI Data as Temporal Hypergraph Brain Network

To capture both temporal and higher-order interactions among brain regions, we represent fMRI data as temporal hypergraphs. The set of Regions of Interest (ROIs), denoted by $\mathcal{V} = \{v_1, \dots v_R\}$, is defined using brain parcellation atlases, with R indicating the number of ROIs based on the atlas. Using the same atlas for all the subjects' fMRI data, the set of ROIs remains identical across individuals; $\mathcal{G}_i = (\mathcal{V}, \mathcal{E}_{\mathcal{G}_i})$ represents the temporal hypergraph of the ith subject.

To capture the dynamic patterns in brain activity, we model the temporal properties of brain networks using the sliding window technique [25]. For the BOLD signals of the ith subject $S_i \in \mathbb{R}^{R \times T}$, with T denoting the total fMRI time interval and $S_i[v_m]$ as the BOLD signal for the mth ROI, we divide S_i into windows $\{S_i^1, \dots, S_i^M\}$, where $M = \lfloor \frac{T-L}{s} \rfloor + 1$, L is the window length, and s is the stride between windows.

To capture higher-order connections and generate the hypergraph $\mathcal{G}_i$, a subset of ROIs $\{u_1, \dots, u_k\} \in \mathcal{V}$ form a hyperedge if the similarities between their corresponding BOLD signals within a windows exceed a threshold. With a similarity measure function, ζ, and $e = \{u_1, \dots, u_k\}$, we have:

$$(e, t_p) \in \mathcal{E}_{\mathcal{G}_i} \text{ if } \zeta(S_i^p[u_1], \dots, S_i^p[u_k]) \geq \text{Threshold}$$

We calculate signal similarity using the Pearson correlation coefficient and form a hyperedge between a set of ROIs where each ROI is in the top 90th percentile of positive correlations of all the other ROIs in the hyperedge.

2.2   Brain Walk Sampling

As a walk-based graph learning approach, we sample a set of random walks over our brain network to automatically extract temporal, higher-order patterns of brain activity. To accommodate the unique characteristics of brain networks, we introduce BRAINWALK. Inspired by SETWALK [3], each BRAINWALK consists of a random sequence of *hyperedges*, allowing us to effectively capture the dynamics of higher-order brain networks. In contrast to many temporal networks, where timestamps represent discrete moments, in our representation, a timestamp represents a continuous interval (window) of BOLD signals, capturing distinct brain states. Considering this property, BRAINWALK uses backward, timestamp-based traversal to capture historical information and intra-timestamp traversal to capture patterns that co-occur within a span of time corresponding to brain activity.

A BRAINWALK with length ℓ on $\mathcal{G} = (\mathcal{V}, \mathcal{E})$ is defined as:

$$BrainWalk : (e_1, t_{e_1}) \rightarrow (e_2, t_{e_2}) \rightarrow \cdots \rightarrow (e_{\ell-1}, t_{e_{\ell-1}}) \rightarrow (e_\ell, t_{e_\ell}),$$

where $e_i \in \mathcal{E}$, and consecutive pairs of e_i and e_{i+1} represent neighboring hyperedges, satisfying the condition $t_{e_i} \geq t_{e_{i+1}}$. The notation $BW[i]$ represents the i-th pair in the walk, where $BW[i][0] = e_i$ and $BW[i][1] = t_{e_i}$.

In our sampling approach, we take into account the temporal proximity of timestamps. This consideration is crucial for understanding the transitions between different states and the lasting effects of the previous task or state of the brain. Therefore, a closer timestamp is likely to be more relevant. To capture this temporal relevance, we use a biased sampling walk. We sample (e, t), a neighboring hyperedge of a previously sampled hyperedge (e_{prev}, t_{prev}) with a probability that scales according to $\exp(\theta(t - t_{prev}))$. Here, θ represents the hyperparameter for the sampling time bias.

2.3 Neural Hyperedge Anomaly Detection

Neural Walk Encoding. Research in graph learning has shown that anonymizing node identities enables models to perform well in an inductive setting and generalizing to unseen patterns by learning general rules unconstrained by specific node identities [3,31]. Following them, to ensure model performance on unseen patterns, we use a two-step anonymization process to conceal hyperedge identities [3]. Initially, we anonymize node identities by replacing them with positional encodings, capturing the occurrence of nodes in different positions across a set of sampled BRAINWALKs. Subsequently, to compute the anonymized encoding of a hyperedge, we aggregate the anonymized node identities corresponding to the nodes it connects. This aggregation is performed using SETMIXER [3], a permutation-invariant pooling strategy based on MLP-MIXER [29]. Finally, we encode each BRAINWALK. Specifically, during the encoding of a BRAINWALK, $\hat{bw}$, we use MLP-MIXER [29] to merge information from the sequence of hyperedge encodings and their corresponding timestamp encodings, resulting in the calculation of $\text{ENC}(\hat{bw})$. For encoding the hyperedge timestamps, we follow previous work on random Fourier features [17] to obtain a vector representation for each timestamp assigned to a hyperedge in the BRAINWALK.

Anomaly Score. To calculate anomaly scores for each hyperedge $e = \{u_1, \ldots, u_k\}$, we use a neural encoding module. This module processes a set of sampled BRAINWALKs starting from each node within the hyperedge. The anomaly score, $\varphi(e)$, is computed as follows:

$$\varphi(e) = \text{MLP}\left(\Psi\left(\{\text{ENC}_{\mathbf{u_1}}, \ldots, \text{ENC}_{\mathbf{u_k}}\}\right)\right) \quad (1)$$

where $\text{ENC}_{u_i} = \frac{1}{N}\sum_{\hat{bw} \in \text{BW}_{u_i}} \text{ENC}\left(\hat{bw}\right)$. Here MLP is a 2-layer perceptron, Ψ is SETMIXER [3] and Bw_{u_i} is the set of N sampled BRAINWALKs. For each walk $\hat{bw} \in \text{Bw}_{u_i}$, it holds that $u_i \in \hat{bw}[0][0] \cap \hat{bw}[1][0]$.

2.4   Training

In the training phase, to be able to learn individual-level as well as group-level patterns common among all subjects, we work with a set of healthy brains. From each brain's fMRI data, we generate the corresponding hypergraph. Adopting the widely employed negative sampling approach [8], we generate a negative sample for every hyperedge $e \in \mathcal{E}$ within the brain hypergraph. This involves keeping 50% of the nodes from e and substituting the remainder with nodes from $\mathcal{V} - e$, resulting in a negative sample. This approach could inadvertently produce a hyperedge, e_j, that is already present in the hypergraph while generating a negative sample for another hyperedge, e_i. To overcome the limitation and ensure the reliability of these negative samples, we introduce a new step not found in prior work [3,31]. We verify that any generated negative hyperedge does not appear in any timestamp of the healthy brain and has never been considered as a normal co-activation in the brain network. Subsequently, we calculate the anomaly score, as defined in Eq. 1, for every hyperedge in the training set, including both normal hyperedges and negative samples. The framework is then trained using a contrastive learning approach. A key advantage of HYPERBRAIN is that it relies solely on data from the healthy control group for training, taking advantage of the abundance of healthy data.

To enhance learning across all subjects' brains in the training data and mitigate the impact of noise inherent to individual networks, we use a two-step approach: pre-training on a subset of healthy brain networks designated for training, followed by fine-tuning on the remaining brain datasets in the training data. This ensures that the model learns from all available brain networks in the training data, promoting a more comprehensive understanding of normal and anomalous patterns across diverse subjects.

3   Evaluation

Our evaluation addresses two questions about the performance of HYPERBRAIN:

1. How effectively does HYPERBRAIN perform in the task of anomalous hyperedge (abnormal co-activation) detection compared to the baselines? (Sect. 3.2)
2. Are the abnormal activities detected by HYPERBRAIN in the brains of people with disorders meaningful and consistent with established research on the disorder's impact on the brain? (Sect. 3.3)

3.1   Datasets and Baselines

We conducted experiments using two real-world fMRI datasets: ① *ADHD data* [5] includes neuroimaging data from 50 subjects diagnosed with ADHD and 50 typically developing (TD) controls. ② *ASD data* [10] contains fMRI data from 45 individuals with Autism and 45 subjects in healthy control group. For brain parcellation, we used the CC200 [11] atlas.

Table 1. Performance in Anomalous Hyperedge Detection: Mean AUC (%). The best result is indicated in boldface.

Datasets \ Methods	Structured-based			Embedding-based			
	CN [23]	JC [24]	AA [35]	PMNE [22]	NHP [33]	CAW-N [21]	HYPERBRAIN
ADHD	75.21	78.00	75.47	75.53	72.12	86.26	**92.33**
ASD	86.47	86.35	86.66	72.54	82.42	83.56	**93.78**

For baselines, we compare HYPERBRAIN with six state-of-the-art approaches in two group of methods: structured-based methods and embedding-based methods. Structured-based Methods are: ① Common Neighbor (CN) [23]: one of the most widely used metrics in anomaly detection based on graph structure. It quantifies overlap or similarity between sets of connections in a network, assuming more normal connections between nodes with a higher number of common neighbors. ② Jaccard Coefficient (JC) [24]: a normalized version of CN. ③ Adamic/Adar (AA) [35]: a weighted version of JC that assigns higher importance to less connected common neighbors. Embedding-based Methods are: ④ Principled Multilayer Network Embedding (PMNE) [22]: a multiplex graph learning method that analyzes functional connectivity by considering each subject as a different type of edge. ⑤ Neural Hypergraph Link Prediction (NHP) [33]: a deep hypergraph learning method that analyzes static brain hypergraphs. ⑥ Causal Anonymous Walks (CAW-N) [21]: a deep learning walk-based temporal graph learning method that analyzes dynamic functional connectivity.

3.2 Quantitative Evaluation on Synthetic Anomalous Hyperedges

To assess HYPERBRAIN's effectiveness in detecting anomalous hyperedges in the brain, we evaluate it on a set of control brain networks belonging to healthy subjects not used in the training phase. Synthetic anomalous hyperedges are injected into these networks following techniques from prior work [8], with further enhancements explained in Sect. 2.4. Subsequently, we deploy HYPERBRAIN to detect these synthetically injected anomalies, assessing performance using the Area Under the ROC Curve (AUC).

The results presented in Table 1 demonstrate that HYPERBRAIN outperforms the baselines by a large margin. Three key factors contribute to HYPERBRAIN's superior performance: ① capturing higher-order patterns, ② incorporating temporal changes in the brain, and ③ the exclusive design and training approach of HYPERBRAIN for brain network considering its unique properties.

3.3 Clinical Relevance and Consistency on Real Datasets

To answer the second question, mirroring real-world scenarios, we investigate anomalies detected by HYPERBRAIN in individuals with specific disorders. Our analysis focuses on assessing whether these detected anomalous hyperedges align with existing research findings related to these disorders. We train HYPERBRAIN

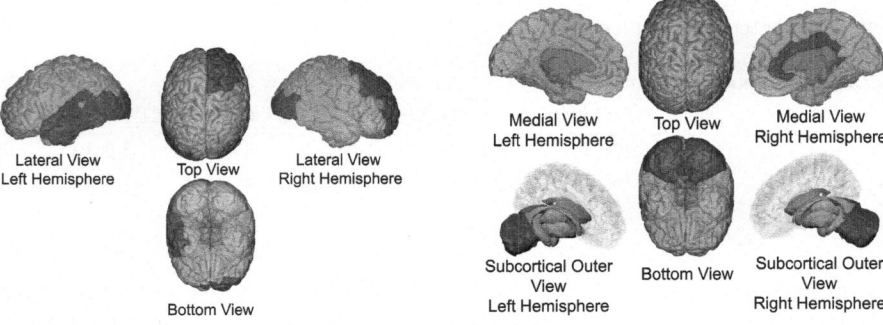

Fig. 2. ADHD-Related Brain Regions Identified by HYPERBRAIN

Fig. 3. ASD-Related Brain Regions Identified by HYPERBRAIN

on neuroimaging data from healthy brains and subsequently test the trained model on brains of individuals diagnosed with disorders. After detecting anomalous hyperedges using HYPERBRAIN, we record the occurrence of each region. Subsequently, we identify and report regions exhibiting statistically significant occurrences in the detected anomalous hyperedges. In the subsequent sections, we analyze identified regions linked to ADHD and ASD.

ADHD. Brain regions with statistically significant occurrences in detected anomalous hyperedges include: *Frontal Pole, Right Frontal Gyrus, Lateral Occipital Cortex*, and *Left Temporal Gyrus* (refer to Fig. 2). Notably, the *Frontal Pole* and *Right Frontal Gyrus*, both located in the *Prefrontal Cortex*, exhibit the highest occurrence rates. The *Prefrontal Cortex* is known for its crucial role in attention regulation and has been associated with impaired function in individuals with ADHD [2]. The other anomalous regions identified in our study are also consistent with prior studies on ADHD [28,34]. Studies reported abnormally low activity in response to a stimulus in the *Left Temporal Gyrus* [34], as well as increased cortical thickness in the *Occipital Cortex* [28], among individuals diagnosed with ADHD.

ASD. Brain regions statistically linked to these anomalous hyperedges include the *Cerebellum* and *Right Cingulate Gyrus* (refer to Fig. 3). Notably, the *Cerebellum*, the most frequently occurring region in detected abnormal hyperedges has been recognized as a key brain area affected in autism [26]. Additionally, the other detected region, the *Right Cingulate Gyrus*, has been reported to exhibit atypical structure and activity in autistic brains [9].

4 Conclusion

We introduce HYPERBRAIN, a novel method to identify biomarkers associated with neuro disorders via identifying anomaly patterns of high-order interactions among brain regions. To this end, we model fMRI data as temporal hypergraphs to effectively capture dynamic higher-order interactions. HYPERBRAIN uses higher-order random walks and a neural encoding to learn intricate patterns in the network. Trained on a set of healthy brain networks, it identifies anomalous patterns in the brain of individuals with disorders. Our evaluation shows that ① HYPERBRAIN outperforms other baselines in hyperedge anomaly detection, and ② the detected abnormal brain activities align consistently with clinical research on disorders. These results suggest several promising directions for future research, including deeper exploration of brain networks, enhanced understanding of symptoms, and early disorder detection. Additionally, as computational cost is a common challenge in hypergraph analysis, another direction for future work is improving the computational efficiency of our method, where the cost is $O(\#subjects \times \#temporal\_hyperedges)$.

Acknowledgments and Disclosure of Funding. We acknowledge the support of the Natural Sciences and Engineering Research Council of Canada (NSERC).

Nous remercions le Conseil de recherches en sciences naturelles et en génie du Canada (CRSNG) de son soutien.

References

1. Abraham, A., et al.: Deriving reproducible biomarkers from multi-site resting-state data: an autism-based example. Neuroimage **147**, 736–745 (2017)
2. Arnsten, A.F.: The emerging neurobiology of attention deficit hyperactivity disorder: the key role of the prefrontal association cortex. J. Pediatr. **154**(5), I (2009)
3. Behrouz, A., Hashemi, F., Sadeghian, S., Seltzer, M.: CAt-walk: inductive hypergraph learning via set walks. Adv. Neural Inf. Process. Syst. **36** (2024)
4. Behrouz, A., Seltzer, M.: ADMIRE++: explainable anomaly detection in the human brain via inductive learning on temporal multiplex networks. In: ICML 3rd Workshop on Interpretable Machine Learning in Healthcare (IMLH) (2023)
5. Brown, J.A., Rudie, J.D., Bandrowski, A., Van Horn, J.D., Bookheimer, S.Y.: The UCLA multimodal connectivity database: a web-based platform for brain connectivity matrix sharing and analysis. Front. Neuroinform. **6**, 28 (2012)
6. Chang, C., Glover, G.H.: Time-frequency dynamics of resting-state brain connectivity measured with fMRI. Neuroimage **50**(1), 81–98 (2010)
7. Chatterjee, T., Albert, R., Thapliyal, S., Azarhooshang, N., DasGupta, B.: Detecting network anomalies using Forman-Ricci curvature and a case study for human brain networks. Sci. Rep. **11**(1), 8121 (2021)
8. Chen, C., Liu, Y.Y.: A survey on hyperlink prediction. IEEE Trans. Neural Netw. Learn. Syst. **35**, 15034–15050 (2023)
9. Chien, Y.L., Chen, Y.C., Gau, S.S.F.: Altered cingulate structures and the associations with social awareness deficits and CNTNAP2 gene in autism spectrum disorder. NeuroImage Clin. **31**, 102729 (2021)
10. Craddock, C., et al.: The neuro bureau preprocessing initiative: open sharing of preprocessed neuroimaging data and derivatives. Front. Neuroinform. **7**(27), 5 (2013)
11. Craddock, R.C., et al.: A whole brain fMRI atlas generated via spatially constrained spectral clustering. Hum. Brain Mapp. **33**(8), 1914–1928 (2012)
12. El-Gazzar, A., Thomas, R.M., van Wingen, G.: Dynamic adaptive spatio-temporal graph convolution for fMRI modelling. In: Abdulkadir, A., et al. (eds.) MLCN 2021. LNCS, vol. 13001, pp. 125–134. Springer, Cham (2021). https://doi.org/10.1007/978-3-030-87586-2_13

13. Eslami, T., Almuqhim, F., Raiker, J.S., Saeed, F.: Machine learning methods for diagnosing autism spectrum disorder and attention-deficit/hyperactivity disorder using functional and structural MRI: a survey. Front. Neuroinform. **14**, 62 (2021)
14. Fan, Y.S., et al.: Individual-specific functional connectome biomarkers predict schizophrenia positive symptoms during adolescent brain maturation. Hum. Brain Mapp. **42**(5), 1475–1484 (2021)
15. Gonzalez-Castillo, J., Bandettini, P.A.: Task-based dynamic functional connectivity: recent findings and open questions. Neuroimage **180**, 526–533 (2018)
16. Hutchison, R.M., et al.: Dynamic functional connectivity: promise, issues, and interpretations. Neuroimage **80**, 360–378 (2013)
17. Kazemi, S.M., et al.: Time2Vec: learning a vector representation of time. arXiv preprint arXiv:1907.05321 (2019)
18. Lee, M.H., Smyser, C.D., Shimony, J.S.: Resting-state fMRI: a review of methods and clinical applications. Am. J. Neuroradiol. **34**(10), 1866–1872 (2013)
19. Li, X., et al.: BrainGNN: interpretable brain graph neural network for fMRI analysis. Med. Image Anal. **74**, 102233 (2021)
20. Li, Y., et al.: Construction and multiple feature classification based on a high-order functional hypernetwork on fMRI data. Front. Neurosci. **16**, 848363 (2022)
21. Liu, M., Liu, Y.: Inductive representation learning in temporal networks via mining neighborhood and community influences. In: Proceedings of the 44th International ACM SIGIR Conference on Research and Development in Information Retrieval, pp. 2202–2206 (2021)
22. Liu, W., Chen, P.Y., Yeung, S., Suzumura, T., Chen, L.: Principled multilayer network embedding. In: 2017 IEEE International Conference on Data Mining Workshops (ICDMW), pp. 134–141. IEEE (2017)
23. Newman, M.E.: Clustering and preferential attachment in growing networks. Phys. Rev. E **64**(2), 025102 (2001)
24. Papadimitriou, P., Dasdan, A., Garcia-Molina, H.: Web graph similarity for anomaly detection. J. Internet Serv. Appl. **1**, 19–30 (2010)
25. Peng, L., Wang, N., Xu, J., Zhu, X., Li, X.: GATE: graph CCA for temporal self-supervised learning for label-efficient fMRI analysis. IEEE Trans. Med. Imaging **42**(2), 391–402 (2022)
26. Rogers, T.D., et al.: Is autism a disease of the cerebellum? An integration of clinical and pre-clinical research. Front. Syst. Neurosci. **7**, 15 (2013)
27. Santoro, A., Battiston, F., Petri, G., Amico, E.: Higher-order organization of multivariate time series. Nat. Phys. **19**(2), 221–229 (2023)
28. Sörös, P., et al.: Inattention predicts increased thickness of left occipital cortex in men with attention-deficit/hyperactivity disorder. Front. Psych. **8**, 170 (2017)
29. Tolstikhin, I.O., et al.: MLP-mixer: an all-MLP architecture for vision. Adv. Neural. Inf. Process. Syst. **34**, 24261–24272 (2021)
30. Wang, L., Li, K., Chen, X., Hu, X.P.: Application of convolutional recurrent neural network for individual recognition based on resting state fMRI data. Front. Neurosci. **13**, 434 (2019)
31. Wang, Y., Chang, Y.Y., Liu, Y., Leskovec, J., Li, P.: Inductive representation learning in temporal networks via causal anonymous walks. arXiv preprint arXiv:2101.05974 (2021)
32. Xiao, L., et al.: Multi-hypergraph learning-based brain functional connectivity analysis in fMRI data. IEEE Trans. Med. Imaging **39**(5), 1746–1758 (2019)
33. Yadati, N., et al.: NHP: neural hypergraph link prediction. In: Proceedings of the 29th ACM International Conference on Information & Knowledge Management, pp. 1705–1714 (2020)

34. Yu, M., et al.: Meta-analysis of structural and functional alterations of brain in patients with attention-deficit/hyperactivity disorder. Front. Psych. **13**, 1070142 (2023)
35. Zhou, T., Lü, L., Zhang, Y.C.: Predicting missing links via local information. Eur. Phys. J. B **71**, 623–630 (2009)
36. Zu, C., et al.: Identifying high order brain connectome biomarkers via learning on hypergraph. In: Wang, L., Adeli, E., Wang, Q., Shi, Y., Suk, H.-I. (eds.) MLMI 2016. LNCS, vol. 10019, pp. 1–9. Springer, Cham (2016). https://doi.org/10.1007/978-3-319-47157-0_1

SpaRG: Sparsely Reconstructed Graphs for Generalizable fMRI Analysis

Camila González$^{(\boxtimes)}$ (ID), Yanis Miraoui, Yiran Fan, Ehsan Adeli, and Kilian M. Pohl

Stanford University, Stanford, CA 94305, USA
{camgonza,ymiraoui}@stanford.edu

Abstract. Deep learning can help uncover patterns in resting-state functional Magnetic Resonance Imaging (rs-fMRI) associated with psychiatric disorders and personal traits. Yet the problem of interpreting deep learning findings is rarely more evident than in fMRI analyses, as the data is sensitive to scanning effects and inherently difficult to visualize. We propose a simple approach to mitigate these challenges grounded on sparsification and self-supervision. Instead of extracting post-hoc feature attributions to uncover functional connections that are important to the target task, we identify a small subset of highly informative connections during training and occlude the rest. To this end, we jointly train a (1) sparse input mask, (2) variational autoencoder (VAE), and (3) downstream classifier in an end-to-end fashion. While we need a portion of labeled samples to train the classifier, we optimize the sparse mask and VAE with unlabeled data from additional acquisition sites, retaining only the input features that generalize well. We evaluate our method – **Spa**rsely **R**econstructed **G**raphs (**SpaRG**) – on the public ABIDE dataset for the task of sex classification, training with labeled cases from 18 sites and adapting the model to two additional out-of-distribution sites with a portion of unlabeled samples. For a relatively coarse parcellation (64 regions), SpaRG utilizes only 1% of the original connections while improving the classification accuracy across domains. Our code can be found at www.github.com/yanismiraoui/SpaRG.

Keywords: fMRI · sparsification · domain generalization

1 Introduction

Resting-state functional Magnetic Resonance Imaging (rs-fMRI) has yielded valuable insights into the neural underpinnings of psychiatric disorders and individual traits, facilitating a deeper understanding of shared brain activity patterns among affected individuals [28]. Yet fMRIs, which comprise hundreds of volumes per scan at a low spatial resolution, are difficult for humans to interpret. The

---

C. González and Y. Miraoui—These authors had an equal contribution.

© The Author(s), under exclusive license to Springer Nature Switzerland AG 2025
D. R. Bathula et al. (Eds.): MLCN 2024, LNCS 15266, pp. 46–56, 2025.
https://doi.org/10.1007/978-3-031-78761-4_5

preferred way to analyze functional connectomes is via two-dimensional matrices depicting the correlation of Blood Oxygen Level Dependent (BOLD) signals between brain regions during the scanning period [3]. While this significantly eases interpretation, it still requires reading the connections between dozens to hundreds of brain regions. Selecting an appropriate parcellation granularity that is sufficiently precise to capture the relevant signal yet simple enough to uncover neural underpinnings and prevent model overfitting is hence critical [5, 21].

Deep learning models have achieved state-of-the-art results in detecting personal characteristics from rs-fMRIs at the subject level [4, 9, 13, 14]. Coupled with interpretability methods, such as ROI-selection pooling layers [14], these models can uncover brain regions and connections that are highly indicative of the target. Graph Attention Networks (GATs) have also emerged as a strategy to identify informative features by leveraging the self-attention mechanism of transformers [17, 25]. However, feature attribution and attention values are continuous and can vary widely between predictions. While these strategies provide individual-level model explanations, they do not reduce the number of functional connections considered by the model and are, therefore, often difficult to interpret. Identifying connections that generalize to unseen domains is even more challenging [11, 23]. In this work, we take a different approach from calculating attributions post-hoc by *learning a small set of generalizable neural connections and guaranteeing that all predictions emerge solely from this small feature set.*

We propose **Sparsely Reconstructed Graphs (SpaRG)**, an end-to-end method that jointly trains a sparse input mask, a self-supervised variational autoencoder, and a classifier (Fig. 1). During training, we sparsify the rs-fMRI correlation matrices by multiplying them with a mask $\mathcal{M}$. The sparse input $x' = \mathcal{M} \odot x$ is reconstructed by a variational autoencoder (VAE). The reconstructed functional connectomes are then the input of a Graph Convolutional Network (GCN), which predicts the outcome. As the sparsification and VAE objectives require no ground truth labels, they can be optimized with data from unlabeled sites. This encourages the sparse mask to occlude connections that are susceptible to the acquisition shift, as these comprise a large reconstruction error. Meanwhile, the supervised classification loss training the GCN preserves connections that are informative to the classification objective.

We evaluate our method on the task of sex classification from rs-fMRIs for the public *ABIDE* [6] dataset and explore two levels of atlas granularity, namely the 64- and 1024-dimensional *Dictionaries of Functional Modes (DiFuMo)* [5], which were trained on millions of fMRI volumes acquired over 27 studies. Our empirical results confirm that learning a mask and unsupervised model jointly results in a set of functional connections that are informative for downstream classification and robust across acquisition sites. In fact, SpaRG can retain and *even improve* classification accuracy despite acquisition differences while occluding up to 99% of the connectomes. The resulting feature sets are consistent across validation folds and parcellation schemes and highlight connections previously identified as relevant for sex classification in the literature.

2   Related Work

Previous work supports the benefits of sparsification for countering the curse of dimensionality in fMRI analyses. For example, masking the 70% lowest correlations has resulted in improved detection accuracy of brain disorders from rs-fMRI [26]. Popular regularizers for reducing the feature space during training include *Lasso* [22], *ElasticNet* [29], *Frobenius* [10] and the *k-support Norm* [2,8]. Sparsification can also increase the consistency of connectivity patterns across individuals [19]. Other methods take into account the correlation between predictors [18] or patterns that arise from different fMRI tasks [20].

Similar to our approach, Ahmadi et al. [1] utilize a sparse autoencoder and thresholding to identify relevant connections for Alzheimer's Disease diagnosis. Other self-supervised approaches have been used for pre-training an encoder on fMRI data [15] and extracting subject-specific functional modes from raw fMRIs [12]. For instance, Zhao et al. [27] leverage a VAE for clustering connectivity patterns in dynamic connectome analysis and outlier detection.

We are, to our knowledge, the first to propose an end-to-end semi-supervised sparsification process operating directly on correlation matrices. Our method makes no assumptions about the data-generating process and leverages unlabeled samples, resulting in robust and interpretable downstream classifiers.

3   Methodology

In the following, we outline our learning scenario and the key components of our method, which are visualized in Fig. 1.

After processing the fMRIs, registering them to a common atlas, and clustering voxels into k parcels, we calculate the Pearson correlation between pairwise time courses to obtain matrices of the form $\mathbf{x} \in \mathbb{R}^{k \times k}$. In our setting, labeled data is only available from a subset of sites but we have access to some unlabeled train cases from all sites – a common scenario when performing domain adaptation. We thus have *two training sets* originating from different distributions: a labeled set $\mathcal{D}_{\mathcal{L}}$ with n input-label pairs for our classification objective $\mathcal{D}_{\mathcal{L}} = \{(\mathbf{x}_1, \mathbf{y}_1), \ldots, (\mathbf{x}_n, \mathbf{y}_n)\}$, and a second set, smaller, set $\mathcal{D}_{\overline{\mathcal{L}}}$ containing only m correlation matrices $\mathcal{D}_{\overline{\mathcal{L}}} = \{\mathbf{x}_1, \ldots, \mathbf{x}_m\}$.

Our Goal is Two-Fold: we wish to **(a)** make accurate predictions $\hat{\mathbf{y}}$, generalizing well across acquisition conditions and **(b)** learn a sparse mask $\mathcal{M}$ that highlights a subset of features highly relevant for our task. Our process optimizes three objectives: *sparsification, reconstruction, and classification.*

3.1   Sparsification: $\mathbf{x} \rightarrow \mathbf{x}'$

Central to our approach is the **trainable sparse mask** $\mathcal{M}$. During the learning process, $\mathcal{M} \in \mathbb{R}^{k \times k}$ has real-valued entries $m_{i,j} = [0, 1]$. After training, we binarize $\mathcal{M}$ based on whether $m_{i,j} > \theta$ for a threshold θ based on the percentage

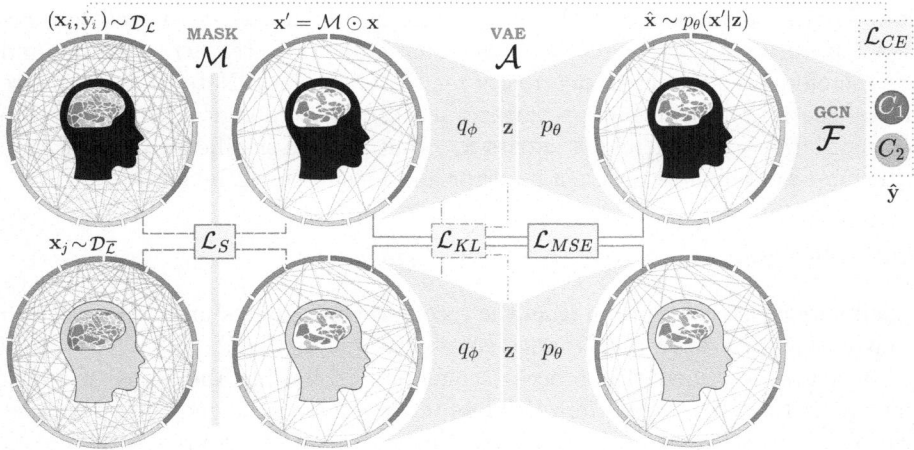

Fig. 1. ***SpaRG:*** a sparse mask $\mathcal{M}$, variational autoencoder (VAE) $\mathcal{A}$ that reconstructs the sparse inputs, and graph convolutional network (GCN) classifier $\mathcal{F}$ are trained in an end-to-end fashion to learn a subset of robust and informative functional connections. We interleave supervised training of the GCN with self-supervised steps, where we optimize the sparsification and autoencoding losses.

of matrix entries to occlude. For encouraging sparsity in $\mathcal{M}$ during training we utilize *ElasticNet* (Eq. 1), which combines Lasso and Ridge penalties.

$$\mathcal{L}_S = \lambda \sum_{i,j} |m_{i,j}| + \frac{1-\lambda}{2} \sum_{i,j} m_{i,j}^2 \qquad (1)$$

Applying the Hadamard product between each input $\mathbf{x}$ and $\mathcal{M}$ results in a *sparsified correlation matrix* $\mathbf{x}' = \mathbf{x} \odot \mathcal{M}$. Note that this operation occurs *in the first step of the forward pass* (see Fig. 1).

3.2   Reconstruction: $\mathbf{x}' \rightarrow \hat{\mathbf{x}}$

Utilizing inputs $\mathbf{x}$ from both $\mathcal{D}_{\mathcal{L}}$ and $\mathcal{D}_{\overline{\mathcal{L}}}$, we learn to *reconstruct the sparse correlation matrix* $\mathbf{x}'$ *into* $\hat{\mathbf{x}}$ with a variational autoencoder $\mathcal{A}$. Our objective here is to minimize the reconstruction $\mathcal{L}_{MSE}$, as well as the Kullback-Leibler (KL) divergence that encourages the prior distribution of the latent space $\mathbf{z}$ to follow a standard normal distribution $\mathbf{z} \sim \mathcal{N}(0,1)$.

$$\mathcal{L}_{MSE} = \frac{1}{n+m} \sum_{i=1}^{n+m} \|\mathbf{x}'_i - \hat{\mathbf{x}}_i\|_2^2; \quad \mathcal{L}_{KL} = \mathrm{KL}\left[q_\phi(\mathbf{z}|\mathbf{x}) \,\|\, \mathcal{N}(\mathbf{0},\mathbf{I}) \right] \qquad (2)$$

This step pursues two objectives. First, by learning a structured latent space with cases from labeled and unlabeled sites, we encourage the autoencoder to learn the same posterior distribution $q(\mathbf{z}|\mathbf{x})$ and likelihood $p(\mathbf{x}|\mathbf{z})$ to reconstruct data from all domains. Second, we teach the sparse mask $\mathcal{M}$ to *occlude functional connections that comprise significant differences between domains* and are therefore reconstructed incorrectly for the OOD data. Note that, as we are reconstructing the sparse input, features masked by $\mathcal{M}$ do not contribute to the reconstruction loss. Therefore, entries $\mathbf{x}_{i,j}$ that diverge significantly across sites will comprise a high reconstruction error and be subsequently occluded.

3.3   Classification: $\hat{\mathbf{x}} \rightarrow \hat{\mathbf{y}}$

Finally, we construct a graph from the reconstructed input $\hat{\mathbf{x}}$ and train a Graph Convolutional Network (GCN) with cross-entropy loss $\mathcal{L}_{CE}$.

We have described this process sequentially following the steps of a forward pass. However, we minimize all loss terms jointly in an end-to-end manner (Eq. 3). Specifically, we perform one training step with $\mathcal{D}_{\mathcal{L}}$ and one with $\mathcal{D}_{\overline{\mathcal{L}}}$. In the second case, we set the classification loss $\mathcal{L}_{CE}$ to zero.

$$\mathcal{L} = \lambda_1 \mathcal{L}_S + \lambda_2 \mathcal{L}_{MSE} + \lambda_3 \mathcal{L}_{KL} + \lambda_4 \mathcal{L}_{CE} \tag{3}$$

By minimizing the joint loss, we learn a sparsification $\mathcal{M}$ that only preserves a fraction of functional connections $\mathbf{x}_{i,j}$ that are informative for our objective.

4   Experimental Setup

4.1   Dataset and Data Preparation

We evaluate SpaRG on the public *Autism Brain Imaging Data Exchange (ABIDE)* [6] dataset, which provides a rich basis for comparison with established baselines. The data comprises rs-fMRIs from individuals with autism spectrum disorder and healthy controls acquired at 20 sites. Our in-distribution (ID) data (F: 50, M: 386; 17.87 ± 8.29) consists of controls without autism spectrum disorder from 18 sites. Cases from sites KKI and NYU form our out-of-distribution (OOD) dataset (F: 50, M: 189; 14.02 ± 6.20), which differs from the ID data in terms of acquisition site, age, and sex distribution. We perform five-fold cross-validation, training on each run with 80% of the ID train data (the rest is used for setting hyperparameters) and 20% of the OOD data. We do not utilize the annotations for the 20% OOD data, simulating a setting where only a few unlabeled cases are available from the target domain. We report the balanced accuracy on ID test data and the remaining 80% cases from KKI &NYU.

For obtaining connectivity matrices, we apply the *Dictionaries of Functional Modes (DiFuMo)* [5], which define 64 or 1024 *soft* brain regions capturing population-wise and individual dynamics. Dadi et al. [5] specify, for each region, which network from the 17-network atlas by Yeo at al. [24] the region belongs to. This allows us to compare the connectivity patterns identified by the models trained with different parcellations.

Table 1. Balanced accuracy, averaged over 5 cross-validation folds, for the task of sex classification on the ABIDE dataset using multiple sparsification strategies and two different parcellation granularities: 64×64 (left) and 1024×1024 (right).

	DiFuMo 64×64			DiFuMo 1024×1024		
	ID	OOD	$|\mathcal{M}|$	ID	OOD	$|\mathcal{M}|$
GCN	76.17 ± 2.2	71.77	.00	77.24 ± 2.7	81.82	.00
FCN	73.94 ± 4.2	61.24	.00	78.34 ± 2.7	80.45	.00
xGW-GAT [17]	46.89 ± 8.2	29.19	.00	40.12 ± 3.5	43.06	.00
Mask-GCN [26]	76.14 ± 2.6	71.17	.70	76.83 ± 3.4	72.40	.70
LASSO [22]	82.10 ± 8.4	14.83	.99	83.74 ± 2.4	**84.69**	**.90**
ElasticNet [29]	76.97 ± 2.8	72.73	.00	83.55 ± 2.1	82.76	$.80$
Frobenius [10]	74.24 ± 5.9	56.94	.00	82.55 ± 5.0	83.25	$.80$
SpaRG (ours)	**82.40 ± 4.5**	**85.17**	**.99**	**84.28 ± 5.5**	82.77	$.80$

4.2 Model Architectures and Baselines

SpaRG is composed of a VAE followed by a GCN. The VAE consists of an encoder with two 16-unit hidden layers and a decoder that mirrors this structure in reverse to reconstruct the input. The classifier has 2 GCN and 2 MLP layers, each comprising 2 units. We train models with *Adam* and a learning rate of 3e−4 until convergence. Given their small size, all models can be trained in a CPU.

We compare SpaRG to multiple baselines and ablations. Alternative sparsification strategies include masking the lowest correlations (*Mask-GCN*) [26], *LASSO* sparsification [22], *ElasticNet* [29] and the *Frobenius* norm [10]. We also compare our GCN-based classifier to the *explainable, geometric, weighted-graph attention network (xGW-GAT)* [17]. Finally, we report ablation results of using only labeled data (SpaRG $\mathcal{D_L}$), not utilizing any sparsification or masking (SpaRG $\overline{\mathcal{L}_S}$) and using a regular autoencoder instead of a VAE (SpaRG AE). We select hyperparameters for all methods via grid search with a validation set consisting of 20% of the ID data. These comprise the weights of the sparsification, autoencoding, and classification terms $\lambda_i \in [0.1, 0.25, 0.5]$ and the mask binarization threshold $|\mathcal{M}| \in [0, 0.7, 0.8, 0.9, 0.95, 0.98, 0.99]$, which determines the ratio of lowest correlations to fully occlude after training.

5 Results

We start by exploring whether we can obtain a small, informative subset of brain connections that permit accurate downstream classification and compare SpaRG to existing strategies. We then conduct an ablation study where we empirically confirm that all components in our method are needed. Finally, we make a visual inspection of the functional connections selected by our method for both atlases.

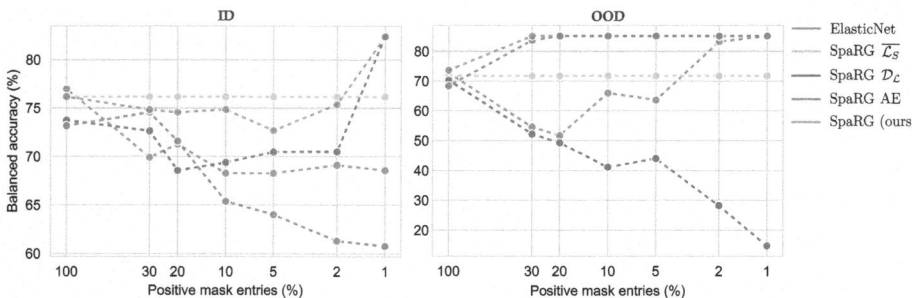

Fig. 2. Balanced accuracy on the ID and OOD sites for different levels of occlusion.

5.1   The Role of Sparsification in Classification Accuracy

In Table 1, we compare the balanced accuracy of our base GCN model (top) with multiple classifier architectures and sparsification strategies for two atlas granularities: 64×64 and 1024×1024. We train only with the controls of ID sites and 20% of the KKI and NYU data as auxiliary OOD unlabeled samples. Before delving into sparsification, we compare three deep learning architectures, namely a GCN, a GAT, and a 4-layered fully connected network (FCN). The GCN obtains the best results, so we proceed with this model as our choice of classifier.

With respect to sparsification, when we utilize the course 64×64 parcellation (left side of the table), most approaches improve classification accuracy on ID data. This supports previous findings on the effectiveness of sparsification to counter the curse of dimensionality in fMRI analysis [26]. However, this only translates to higher OOD accuracy for SpaRG, which leverages a small subset of unlabeled scans from OOD sites. In column $|\mathcal{M}|$, we report the best ratio of occluded connections for each approach, selected on ID validation data. Those connections are occluded after training. Only Mask-GCN, Lasso and SpaRG perform best when occluding a large portion of the connections. The fine-grained 1024×1024 parcellation strategy (right side) is less susceptible to acquisition changes, as reflected in the higher accuracy on OOD data for all methods. This is potentially due to the fine-grained functional modes being more noisy and distinct between individuals [5], preventing the network from overfitting to scanning peculiarities during the training process. In general, utilizing the higher-dimensional matrices coupled with sparsification and post-training occlusion obtains the most reliable results.

5.2   Self-supervision Promotes Generalizable Occlusion

Table 2 summarizes our ablation study of SpaRG. First, we explore a variant that does not perform any sparsification or masking (SpaRG $\overline{\mathcal{L}_S}$). In this setting, the VAE alone does not alleviate the effect of the distribution shift, as shown in the low accuracies for OOD data. We further demonstrate that using unlabeled data

Table 2. Ablative testing of the different components making up SpaRG.

	DiFuMo 64×64			DiFuMo 1024×1024		
	ID	OOD	$|\mathcal{M}|$	ID	OOD	$|\mathcal{M}|$
SpaRG $\overline{\mathcal{L}_S}$	76.17 ± 4.4	71.77	$.00$	83.20 ± 1.4	29.19	**.99**
SpaRG $\mathcal{D}_\mathcal{L}$	$\mathbf{82.40 \pm 4.5}$	14.83	**.99**	84.02 ± 2.4	85.16	**.99**
SpaRG AE	74.55 ± 4.1	83.73	$.70$	84.01 ± 4.3	**85.17**	$.98$
SpaRG (ours)	$\mathbf{82.40 \pm 4.5}$	$\mathbf{85.17}$	**.99**	$\mathbf{84.28 \pm 5.5}$	82.77	$.80$

improves generalization as opposed to only leveraging labeled ID cases (SpaRG $\mathcal{D}_\mathcal{L}$). Finally, we establish that a VAE – which shapes the latent space to follow a standard normal – is preferable over a regular autoencoder (SpaRG AE).

Beyond finding a solution for a specific occlusion threshold, we conduct an analysis of multiple specification options for the 64×64 atlas. Figure 2 corroborates that SpaRG, grounded in self-supervised reconstruction, helps guide the sparsification for multiple thresholds.

5.3 Qualitative Examination of the Preserved Functional Connections

Figure 3 shows the connections preserved by SpaRG for models trained with both parcellation granularities, clustered for comparison purposes into the networks of the Yeo et al. [24] atlas following Dadi et al. [5]. A visual inspection of the connectivity between networks demonstrates that similar patterns are learned by both models. Evidently, for classifying the sex from rs-fMRI, the models

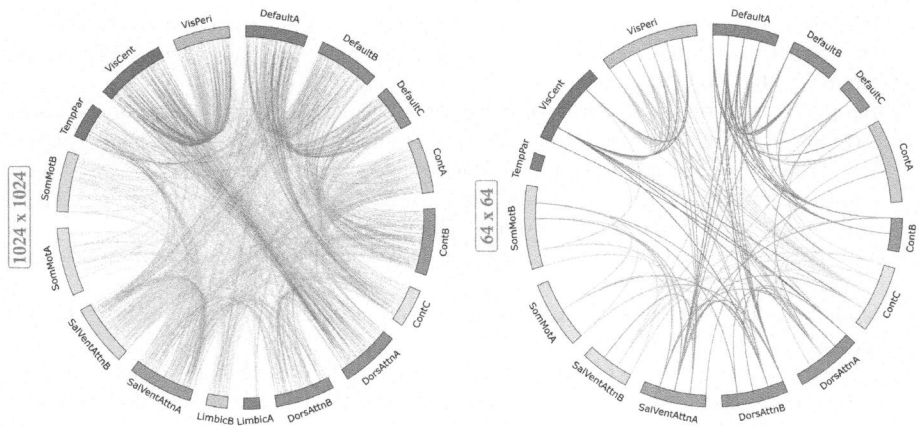

Fig. 3. Functional connections preserved by *SpaRG* for two parcel granularities, mapped to the 17-network atlas [24]. Similar connections are preserved by both models, highlighting connectivity involving the visual and default mode networks.

utilize connections that implicate visual and attention functions and the default mode network, supporting previous findings [7,16]. In this work, we focused on the well-understood task of sex classification, which allowed us to examine the potential and limitations of SpaRG beyond domain-specific design choices. Our results indicate that self-supervised sparsification can potentially allow a better exploration of the underlying mechanisms of psychiatric disorders, as we will explore in additional settings in future work.

6   Conclusion

Functional MRI connectivity data holds immense potential for advancing the understanding of psychiatric and neurodegenerative disorders. Yet the intrinsic difficulty in interpreting high-dimensional correlation matrices and the small reproducibility of findings across acquisition sites and populations introduce significant hurdles. We propose an alternative avenue to observing subject-level feature attributions, namely learning a sparse mask that occludes uninformative functional connections alongside a VAE that identifies connections stable across distribution shifts through self-supervision. Optimizing these components and a downstream classifier jointly allows us to find a subset of up to 1% the size of the original correlation matrices while preserving or improving classification accuracy. These findings highlight the potential of self-supervised sparsification for increasing the interpretability of fMRI analyses.

Acknowledgement. The work was partly funded by the U.S. National Institutes of Health (NIH) grants (DA057567), Stanford HAI Google Cloud Credit, the DGIST Joint Research Project, and the 2024 Stanford HAI Hoffman-Yee Grant.

References

1. Ahmadi, H., Fatemizadeh, E., Motie-Nasrabadi, A.: Deep sparse graph functional connectivity analysis in AD patients using fMRI data. Comput. Methods Programs Biomed. **201**, 105954 (2021)
2. Argyriou, A., Foygel, R., Srebro, N.: Sparse prediction with the k-support norm. Adv. Neural Inf. Process. Syst. **25** (2012). www.proceedings.neurips.cc/paper/2012/hash/99bcfcd754a98ce89cb86f73acc04645-Abstract.html
3. Buxton, R.B.: Introduction to Functional Magnetic Resonance Imaging: Principles and Techniques. Cambridge University Press, San Diego (2009)
4. Cui, H.: BrainGB: a benchmark for brain network analysis with graph neural networks. IEEE Trans. Med. Imaging **42**(2), 493–506 (2022)
5. Dadi, K., et al.: Fine-grain atlases of functional modes for fMRI analysis. Neuroimage **221**, 117126 (2020)
6. Di Martino, A., et al.: The autism brain imaging data exchange: towards a large-scale evaluation of the intrinsic brain architecture in autism. Mol. Psychiatry **19**(6), 659–667 (2014)
7. Gadgil, S., Zhao, Q., Pfefferbaum, A., Sullivan, E.V., Adeli, E., Pohl, K.M.: Spatio-temporal graph convolution for resting-state fMRI analysis. In: Martel, A.L., et al. (eds.) MICCAI 2020. LNCS, vol. 12267, pp. 528–538. Springer, Cham (2020). https://doi.org/10.1007/978-3-030-59728-3_52

8. Gkirtzou, K., Honorio, J., Samaras, D., Goldstein, R., Blaschko, M.B.: fMRI analysis of cocaine addiction using k-support sparsity. In: IEEE 10th International Symposium on Biomedical Imaging, pp. 1078–1081. IEEE (2013)
9. Kan, X., Cui, H., Lukemire, J., Guo, Y., Yang, C.: FBNetGen: task-aware GNN-based fMRI analysis via functional brain network generation. In: International Conference on Medical Imaging with Deep Learning, pp. 618–637 (2022)
10. Krauthgamer, R., Sapir, S.: Comparison of matrix norm sparsification. Algorithmica **85**(12), 3957–3972 (2023)
11. Lee, J., Kang, E., Jeon, E., Suk, H.-I.: Meta-modulation network for domain generalization in multi-site fMRI classification. In: de Bruijne, M., et al. (eds.) MICCAI 2021. LNCS, vol. 12905, pp. 500–509. Springer, Cham (2021). https://doi.org/10.1007/978-3-030-87240-3_48
12. Li, H., et al.: Computing personalized brain functional networks from fMRI using self-supervised deep learning. Med. Image Anal. **85**, 102756 (2023)
13. Li, X., Dvornek, N.C., Zhou, Y., Zhuang, J., Ventola, P., Duncan, J.S.: Graph neural network for interpreting task-fMRI biomarkers. In: Shen, D., et al. (eds.) MICCAI 2019. LNCS, vol. 11768, pp. 485–493. Springer, Cham (2019). https://doi.org/10.1007/978-3-030-32254-0_54
14. Li, X., et al.: BrainGNN: interpretable brain graph neural network for fMRI analysis. Med. Image Anal. **74**, 102233 (2021)
15. Malkiel, I., Rosenman, G., Wolf, L., Hendler, T.: Self-supervised transformers for fMRI representation. In: Proceedings of the 5th International Conference on Medical Imaging with Deep Learning, pp. 895–913. Proceedings of Machine Learning Research (2022)
16. Müller-Oehring, E.M., et al.: Influences of age, sex, and moderate alcohol drinking on the intrinsic functional architecture of adolescent brains. Cereb. Cortex **28**(3), 1049–1063 (2018)
17. Nerrise, F., Zhao, Q., Poston, K.L., Pohl, K.M., Adeli, E.: An explainable geometric-weighted graph attention network for identifying functional networks associated with gait impairment. In: Greenspan, H., et al. (eds.) MICCAI 2023. LNCS, vol. 14221, pp. 723–733. Springer, Cham (2023). https://doi.org/10.1007/978-3-031-43895-0_68
18. Ng, B., Abugharbieh, R.: Generalized sparse regularization with application to fMRI brain decoding. In: Székely, G., Hahn, H.K. (eds.) IPMI 2011. LNCS, vol. 6801, pp. 612–623. Springer, Heidelberg (2011). https://doi.org/10.1007/978-3-642-22092-0_50
19. Ng, B., Varoquaux, G., Poline, J.-B., Thirion, B.: A novel sparse graphical approach for multimodal brain connectivity inference. In: Ayache, N., Delingette, H., Golland, P., Mori, K. (eds.) MICCAI 2012. LNCS, vol. 7510, pp. 707–714. Springer, Heidelberg (2012). https://doi.org/10.1007/978-3-642-33415-3_87
20. Rao, N., Cox, C., Nowak, R., Rogers, T.T.: Sparse overlapping sets lasso for multitask learning and its application to fMRI analysis. Adv. Neural Inf. Process. Syst. **26** (2013). www.proceedings.neurips.cc/paper_files/paper/2013/hash/a1519de5b5d44b31a01de013b9b51a80-Abstract.html
21. Schaefer, A., et al.: Local-global parcellation of the human cerebral cortex from intrinsic functional connectivity MRI. Cereb. Cortex **28**(9), 3095–3114 (2018)
22. Tibshirani, R.: Regression shrinkage and selection via the lasso. J. R. Stat. Soc. Ser. B Stat Methodol. **58**(1), 267–288 (1996)
23. Wang, M., Zhang, D., Huang, J., Yap, P.T., Shen, D., Liu, M.: Identifying autism spectrum disorder with multi-site fMRI via low-rank domain adaptation. IEEE Trans. Med. Imaging **39**(3), 644–655 (2020)

24. Yeo, B.T., et al.: The organization of the human cerebral cortex estimated by intrinsic functional connectivity. J. Neurophysiol. **106**(3), 1125–1165 (2011)

25. Yin, W., Li, L., Wu, F.X.: A graph attention neural network for diagnosing ASD with fMRI data. In: 2021 IEEE International Conference on Bioinformatics and Biomedicine, pp. 1131–1136. IEEE (2021)

26. Zhang, J., Wang, Q., Wang, X., Qiao, L., Liu, M.: Preserving specificity in federated graph learning for fMRI-based neurological disorder identification. Neural Netw. **169**, 584–596 (2024)

27. Zhao, Q., Honnorat, N., Adeli, E., Pfefferbaum, A., Sullivan, E.V., Pohl, K.M.: Variational autoencoder with truncated mixture of Gaussians for functional connectivity analysis. In: Chung, A.C.S., Gee, J.C., Yushkevich, P.A., Bao, S. (eds.) IPMI 2019. LNCS, vol. 11492, pp. 867–879. Springer, Cham (2019). https://doi.org/10.1007/978-3-030-20351-1_68

28. Zhao, Q.: Adolescent alcohol use disrupts functional neurodevelopment in sensation seeking girls. Addict. Biol. **26**(2), e12914 (2021)

29. Zou, H., Hastie, T.: Regularization and variable selection via the elastic net. J. R. Stat. Soc. Ser. B Stat Methodol. **67**(2), 301–320 (2005)

A Lightweight 3D Conditional Diffusion Model for Self-explainable Brain Age Prediction in Adults and Children

Matthias Wilms[1,2,3,4](✉), Ahmad O. Ahsan[1,2,3,4], Erik Y. Ohara[2,3,4],
Gabrielle Dagasso[2,3,4], Elizabeth Macavoy[2,3,4], Emma A. M. Stanley[2,3,4],
Vibujithan Vigneshwaran[2,3,4], and Nils D. Forkert[2,3,4]

[1] Departments of Pediatrics and Community Health Sciences, University of Calgary,
Calgary, Canada
matthias.wilms@ucalgary.ca
[2] Department of Radiology, University of Calgary, Calgary, Canada
[3] Hotchkiss Brain Institute, University of Calgary, Calgary, Canada
[4] Alberta Children's Hospital Research Institute,
University of Calgary, Calgary, Canada

Abstract. Deep learning models that predict an individual's biological brain age from structural MR images are widely used in neuroimaging to analyze brain development and aging. While standard discriminative models for this task are highly accurate, they usually suffer from poor explainability. This can be overcome by inherently self-explainable models that follow a generative paradigm that not only enables them to perform predictions but also to generate counterfactual images that visually explain their decision making process. Lately, denoising diffusion models have achieved incredible results in generative machine learning. However, their training on full-resolution 3D MR images, as required for many tasks, is computationally expensive. Here, we present a new lightweight wavelet-based diffusion model architecture that enables computationally efficient training of age-conditioned diffusion models on 3D brain MR images with a single consumer-grade GPU. We then show how those lightweight diffusion models can be used for self-explainable biological brain age prediction. The results of our proof-of-concept evaluation relying on imaging data of more than 7000 adults and more than 5000 images of children demonstrate the effectiveness of our method. This includes diverse, and realistic samples generated by the models, accurate brain age predictions (mean absolute error for adults: 3.93 yrs., children: 1.38 yrs.), and realistic counterfactual images.

1 Introduction

Discriminative deep neural networks (DDNNs), such as disease classifiers and brain age prediction models, have revolutionized neuroimage analysis by enabling unprecedented diagnostic accuracies and by providing insights into the brain's structure and function using 3D imaging data [32]. This success has largely

© The Author(s), under exclusive license to Springer Nature Switzerland AG 2025
D. R. Bathula et al. (Eds.): MLCN 2024, LNCS 15266, pp. 57–67, 2025.
https://doi.org/10.1007/978-3-031-78761-4_6

been driven by architectural advancements from convolutional neural networks (CNNs) to transformers and beyond and by the increasing availability of vast datasets. However, the clinical adoption of DDNNs in neuroimaging remains limited. A major obstacle is the black box nature of these models, which prevents users from understanding their decision-making processes, thereby creating trust issues [32]. To address this, the community has developed numerous explainability methods [28], most of which are gradient-based, such as saliency maps, and applied post-hoc to trained DDNNs. However, this raises questions about the faithfulness of their explanations. Moreover, saliency maps typically only indicate the spatial source of the decision without explaining why an alternate outcome was not chosen. This can be addressed by generating counterfactuals of the original image that visualize hypothetical 'What-if?'-scenarios that would have led to a different decision by the DDNN. Counterfactual images have been shown to be much more intuitive and user-friendly means of explanation [14]. For DDNNs, counterfactuals can, for example, be generated via post-hoc approaches that involve generative models such as Generative Adversarial Networks [26].

Instead of trying to explain the behaviour of DDNNs with post-hoc approaches, more recently the paradigm of building inherently explainable (self-explainable) deep neural networks has gained traction [25]. While different ways for achieving this goal exist, if built-in capabilities to generate counterfactuals are desired, the use of a single (conditional) generative model to solve the discriminative task and to generate alternate images is a straightforward idea that offers immediate self-explainability [29]. Instead of learning the conditional probability distribution $p(c|\mathbf{x})$ over classes c and images $\mathbf{x}$ directly as done in DDNNs, generative models learn the joint distribution $p(c, \mathbf{x})$ or conditional distribution $p(\mathbf{x}|c)$ and then utilize Bayes' theorem for the discriminative part [3]. Being able to sample from $p(c, \mathbf{x})$ or $p(\mathbf{x}|c)$ then allows to generate counterfactual images.

In medical image analysis, for example, normalizing flow (NF)-based generative models [29] have been employed as self-explainable generative classifiers or regression models due to their ability to provide exact likelihoods as a result of the guaranteed invertibility of the learned mapping between the data distribution and a base distribution. Examples include [29,30] for biological brain age prediction and/or sex classification from 3D T1-weighted magnetic imaging resonance (MRI). However, the invertibility of NF models is only possible because of the use of restricted, specialized architectures that make building such models challenging and computationally expensive or require substantial dimensionality reduction of the imaging data as a pre-processing step, which may degrade fidelity of the generated counterfactuals or discriminative performance [29].

Most recently, denoising diffusion models have emerged as a new powerful class of generative models with impressive generative capabilities [15]. Here, the forward mapping from data distribution to base distribution (*i.e.*, Gaussian noise) is pre-defined while the step-wise backward transformation is carried out by a trained denoising network. While usually not being able to provide exact likelihoods, diffusion models only rely on standard architectural building blocks ubiquitously used in deep learning (*e.g.*, U-Nets [15] as the denoiser). This makes their implementation and training straightforward, although computationally very expensive for large input images such as 3D MRI data of the brain [15].

Lately, the use of trained (conditional) diffusion models for classification purposes has gained attraction in the computer vision and machine learning domains [6,13,17] and initial results in [13] indicate that diffusion-based generative classifiers could be more robust than standard DDNNs while also featuring intrinsic biases that are more akin to those of the human visual system (*e.g.*, shape-focused). Until now, the use of diffusion models for discriminative purposes has been limited in medical image analysis [4] and we are not aware of any work that utilizes full-resolution 3D MRI data of the brain (*e.g.*, 1 mm isotropic voxel size). We hypothesize that one reason for this is the need for substantial GPU resources to train a diffusion model on such data using standard architectures.

The overarching goal of this work is to close this gap via two major contributions: (1) We propose a lightweight wavelet-based diffusion model architecture that enables computationally efficient training of conditional models on full-resolution 3D brain MRI data with a single consumer-grade GPU (here: Nvidia RTX 4090 with 24 GB RAM) in less than 48 h in an end-to-end way. (2) We explore the use of those lightweight diffusion models conditioned on age for self-explainable biological brain age prediction from T1-weighted 3D brain MRI data of adults and children. Here, self-explainability is achieved via counterfactuals generated by the diffusion models.

The key idea of our lightweight architecture is to substantially reduce the spatial size of the full-resolution 3D MRI images that are fed into the diffusion model's denoiser (here: U-Net) without losing information. This allows us to utilize a smaller U-Net that fits into a single consumer-grade GPU for training and inference. Standard techniques to handle high-dimensional data in diffusion models such as learnable autoencoders [16,24] or basic image downsampling or cropping operations [8] result in an information loss and/or require additional training. In contrast, our models rely on a lossless, non-learnable, and efficient wavelet packet transform that decreases spatial size at the expanse of increasing the number of input/output channels of the denoiser. This concept follows recent trends to incorporate wavelet-based elements into deep learning architectures [18,22,31]. This part of our work is also conceptually similar to the concurrent work of Friedrich et al. [10] to build computationally efficient wavelet-based diffusion models for 3D neuroimage data. The major difference is that we propose the use of a two-level wavelet packet transform [20], which allows us to reduce the spatial size of the input images even further (factor 8) than what is possible with the standard wavelet transformation used in [10]. Moreover, [10] builds unconditional diffusion models for image synthesis, while we build conditional diffusion models and use them for self-explainable brain age prediction.

To summarize, in this work, we present lightweight conditional diffusion models for full-resolution 3D MRI data of the brain that can be trained on a single consumer-grade GPU and employ them for brain age prediction. We train and evaluate our models on imaging data of more than 7000 adults from the UK

Biobank [19] as well as on more than 5000 images of children from several publicly available databases such as the ABCD study [5].

2   Methods

We will first briefly summarize (conditional) diffusion models in Sect. 2.1. Then, the proposed incorporation of a two-level wavelet packet transform into the diffusion model setup will be introduced in Sect. 2.2, before we describe how to use those models for brain age prediction in Sect. 2.3.

2.1   Conditional Diffusion Models

Conditional probabilistic diffusion models [11] are latent variable generative models with parameters θ that model the conditional distribution $p_\theta(\mathbf{x}_0|c)$ using a Markov chain setup with Gaussian transitions, a fixed forward process, and a learnable reverse process. In our setting, samples $\mathbf{x}_0 \in \mathbb{R}^{x \times y \times z}$ are T1-weighted 3D MR images and conditions $c \in \mathbb{N}_{\geq 0}$ represent chronological age values rounded to the next full year. Then, the forward process is used to gradually add Gaussian noise to the original image $\mathbf{x}_0$, which generates a sequence of T increasingly noisier latent variables $\mathbf{x}_{1:T} := (\mathbf{x}_1, \ldots, \mathbf{x}_T)$ of the same spatial size. This process can be gradually reversed using learned transitions $p_\theta(\mathbf{x}_{t-1}|\mathbf{x}_t, c) := \mathcal{N}(\mathbf{x}_{t-1}; \mu_\theta(\mathbf{x}_t, t, c), \Sigma_\theta(\mathbf{x}_t, t, c))$ starting from base distribution $p(\mathbf{x}_T)$ (spherical Gaussian) at time step $t = T$, which leads to [11]:

$$p_\theta(\mathbf{x}_0|c) := \int_{\mathbf{x}_{1:T}} p(\mathbf{x}_T) \prod_{t=1}^{T} p_\theta(\mathbf{x}_{t-1}|\mathbf{x}_t, c) \, \mathrm{d}\mathbf{x}_{1:T} \ . \tag{1}$$

The goal of model training is then to learn parameters θ of a neural network that produces an output that parameterizes the mean of the transition operator so that forward and backward processes match. This is usually done by minimizing the variational lower bound of the negative log-likelihood $-\log(p_\theta(\mathbf{x}_0|c))$ [6,11]. Ignoring constant terms and assuming T to be reasonably large ($e.g.$, $T = 1000$), the loss $\mathcal{L}$ to be minimized can be written as [13]

$$\mathcal{L}(\theta) := \mathbb{E}_{\mathbf{x}_0, c, t, \epsilon} \left[w_t \big\| \mathbf{x}_0 - \hat{\mathbf{x}}_\theta(\mathbf{x}_t, t, c) \big\|^2 \right] , \tag{2}$$

with data sample tuple $(\mathbf{x}_0, c)$, time step $t \sim \mathcal{U}(0, T)$, noise $\epsilon \sim \mathcal{N}(\mathbf{0}, \mathbf{I})$, and time step-dependent weight w_t. Due to the specific definition of the forward process (see [11,17]) for more details), latent variable $\mathbf{x}_t$ can be computed in closed form as a linear combination of $\mathbf{x}_0$ and noise ϵ. Then, $\hat{\mathbf{x}}_\theta(\cdot)$ is the neural network-parameterized function with parameters θ that predicts $\mathbf{x}_0$ from $\mathbf{x}_t$.

2.2   Wavelet Packet Transform for Lightweight Denoising

The neural network $\hat{\mathbf{x}}_\theta(\cdot)$ in Eq. 2 denoises $\mathbf{x}_t$ and often a U-Net [15] with modifications such as time step (and condition) embeddings, self-attention layers, and residual connections [11] is used for this. In the context of high-resolution 3D MRI data of the brain (e.g., isometric 1 mm resolution; close to 256 voxels in each direction), the computational challenge is that such a U-Net needs to learn sufficient information at different scales. This requires them to contain many layers that result in a large GPU memory footprint that makes training on consumer-grade hardware challenging. In each layer of the U-Net, the spatial resolution is reduced and the feature dimension/number of output channels is increased. The key idea of our approach to overcome this problem is to directly reduce the spatial size of the full resolution images and to create an initial set of non-trainable feature channels that the U-Net does not have to learn so that we can reduce its depth, resulting in a more lightweight architecture. As we do not want to lose any image information, we opt for a discrete wavelet packet transform (DWPT) [20], which is a lossless, invertible transform. While similar to the discrete wavelet transform (DWT) [20] for single-level decompositions, the DWPT extends the DWT for higher level decompositions. This is done by applying low-pass and high-pass wavelet filters (followed by a downsampling step) not only to the approximation coefficient images as in the DWT but also to the detail coefficients, which generates a full binary tree of coefficient images (see [20] for details). Therefore, and in contrast to the DWT, all similarly sized coefficient images at each level combined fully represent the original image.

More precisely, we propose to perform a two-level DWPT with Haar wavelet filters [20] that decomposes a 3D image $\mathbf{x} \in \mathbb{R}^{x \times y \times z}$ into 64 coefficient images $\tilde{\mathbf{x}}^2 := \{\tilde{\mathbf{x}}^{2,i}\}_{i=1}^{64} \in \mathbb{R}^{\frac{x}{4} \times \frac{y}{4} \times \frac{z}{4}}$ that only have a quarter of the spatial resolution of $\mathbf{x}$ along each image direction. This transformation into the wavelet domain can be efficiently computed and we then treat $\tilde{\mathbf{x}}^2$ as a 64-channel 3D image and use it as the input $\mathbf{x}_0$ to the diffusion model in Sect. 2.1. When sampling from the diffusion model, this step can be reversed by the inverse DWPT to move from the wavelet domain to the image domain. In comparison to [10], using the two-level DWPT instead of a single-level DWT enables us to further reduce the spatial resolution of $\mathbf{x}$ by a factor of two along each image direction and to an overall level that enables training on a standard GPU with 24 GB RAM (40 GB GPU required in [10], number of channels increase by a factor of eight in our work: 8 [10] vs. 64 channels).

2.3   Self-explainable Brain Age Prediction with Diffusion Models

The goal is now to use the DWPT-enhanced conditional diffusion model as a brain age prediction model. For this, we assume that the model was trained on a training set $\mathcal{T} := \{(\mathbf{x}^i, c^i)\}_{i=1}^N$ of N tuples of images and age values with Eq. 2. Let $\mathcal{C} := \{c_j\}_{j=1}^M \subset \mathbb{N}_{\geq 0}$ be the set of M distinct age values in $\mathcal{T}$. Then, given a test image $\mathbf{x}$ with unknown age $\tilde{c}$, we cast the brain age prediction problem as a probabilistic classification problem over $\mathcal{C}$ with a uniform class prior by using Bayes' theorem [6,17]:

$$\tilde{c} = \arg\max_{c_j \in \mathcal{C}} p_\theta(c_j|\mathbf{x}) = \arg\max_{c_j \in \mathcal{C}} p_\theta(\mathbf{x}|c_j) = \arg\max_{c_j \in \mathcal{C}} \log p_\theta(\mathbf{x}|c_j) . \qquad (3)$$

We then replace $\log p_\theta(\mathbf{x}|c_j)$ with the variational lower bound $-\mathbb{E}_{t,\epsilon}\left[w_t\|\mathbf{x} - \hat{\mathbf{x}}_\theta(\mathbf{x}_t, t, c_j)\|^2\right]$ also used for training the diffusion model (see Eq. 2). Following [6,17], we estimate class-specific expectations via Monte Carlo sampling with K tuples (t_l, ϵ_l) with $t_l \sim \mathcal{U}(0,T)$ and $\epsilon_l \sim \mathcal{N}(\mathbf{0},\mathbf{I})$ as $\frac{1}{K}\sum_{l=1}^{K} w_t\|\mathbf{x} - \hat{\mathbf{x}}_\theta(\mathbf{x}_{t_l}, t_l, c_j)\|^2$ with weight $w_t := \exp(-7t)$ (as recommended in [6]). Here, $\mathbf{x}_{t_l}$ is computed from (t_l, ϵ_l) as during training. Having estimated $\tilde{c}$, we are now interested in explaining this result through counterfactual images for other age values to visually answer the question why the diffusion model assigned the highest likelihood to this age and not another value. We do this by first deterministically embedding image $\mathbf{x}$ into the latent space of the diffusion model by utilizing the inverse DDIM scheduler [27] with age $\tilde{c}$. Using the standard DDIM scheduler [27], we then denoise the embedded image with other age values/conditions c_j to produce versions of $\mathbf{x}$ that the model associates with those c_j values. Visually comparing the synthetic counterfactual images to $\mathbf{x}$, reveal the model's decision making process and results in the desired self-explainability.

3   Experiments and Results

In our proof-of-concept evaluation of the proposed lightweight, DWPT-enhanced conditional diffusion model, we focus on three main aspects: 1) Showing its ability to handle and generate full-resolution 3D brain MRI data. 2) Exploration of the accuracy achievable when used for biological brain age prediction in children and adults. 3) Showcasing the model's built-in ability to generate counterfactuals that explain its brain age predictions (self-explainability).

Data: We use two data sets of T1-weighted brain MR images: One to build a model that captures brain aging in adults (adult data) and one that covers a pediatric population (pediatric data). For the adult data, we utilize images of 7110 healthy individuals from the UK Biobank (age range: 46–82 yrs). The pediatric data set is composed of 5207 images from healthy/typically developing individuals (age range: 5–20 yrs) sourced from three original databases: ABCD [5], HBN [1], ABIDE-I/II [7]. The adult/pediatric data sets are randomly split into training, test, and validation sets of 5897/4214, 1157/633, and 656/360 images, respectively, while stratifying for age and source. Standard pre-processing steps are carried out on both data sets, including: Skull-stripping [12] and affine registration [2] to the MNI space [9]. After pre-processing, all images have a spatial size of $192 \times 256 \times 192$ voxels (isotropic 1 mm spacing).

Experiments: We train separate DWPT-enhanced conditional diffusion models for the adult and pediatric data sets with the process outlined in Sects.

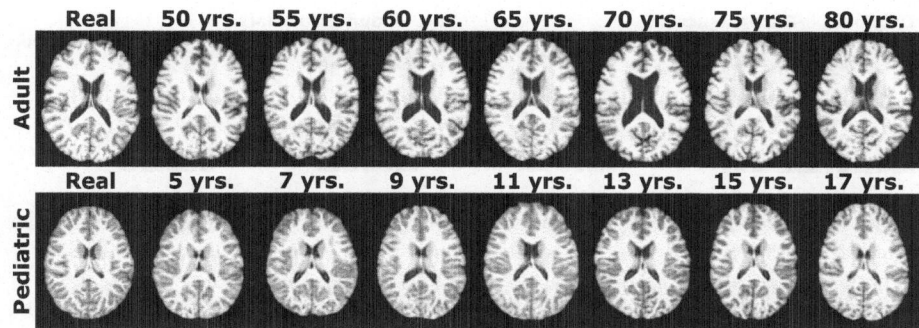

Fig. 1. Axial slices of seven randomly sampled brains from each model (rows) to showcase the general variability captured by the models and their ability to generate realistic full-resolution samples when compared to real images used for training (first column).

2.1/2.2. Both models are conditioned on chronological age (rounded to next full year) resulting in $M = 37/M = 16$ possible conditions or classes for the adult/pediatric model (see Sect. 2.3). Both models use the same U-Net architecture for the denoiser $\hat{\mathbf{x}}_\theta(\cdot)$ consisting of five down-/upsampling blocks (# of output channels: 128, 128, 256, 256, 512) and a total of 232M parameters. Our MONAI-based [23] implementations are available on github[1]. Conditioning is implemented using cross-attention and $T = 1000$ time steps are used during training. We use a standard Nvidia RTX 4090 GPU with 24GM RAM for all experiments. This setup enables us to train $\hat{\mathbf{x}}_\theta(\cdot)$ with a batch size of 5 and we train it for up to 300 epochs with an Adam optimizer (learning rate: 1e−5).

Given the trained models, we sample random synthetic images for different age values with a DDIM scheduler [27] and 100 inference steps. Both models are then also used to estimate the test subjects' brain age using the mechanism described in Sect. 2.3. Here, we use a maximum of $K = 50$ Monte-Carlo trials per class, while dropping the least likely (remaining) half of the ages after each 5 trials to save time. Brain age estimation accuracy is then measured by calculating the mean absolute error (MAE) between chronological and estimated age values. Counterfactual images are generated as described in Sect. 2.3 by noising the real image to $t = 900$ and using 100 inference steps. As a baseline for age estimation accuracy, the commonly used, purely discriminative SFCN model [21] is trained and evaluated for each data set.

Results: Model training converged in under 48 h for both diffusion models (adults: 250 epochs, 44 h, 10:30 min./epoch; pediatric: 300 epochs, 40 h, 8 min./epoch). Despite a slight blurriness, the age-conditioned random samples from both models visualized in Fig. 1 illustrate the general fidelity and quality of the images generated by the models when compared to a real image as well as their realistic morphological diversity. From the samples of the adult model (see also Fig. 2), it can also be seen that the model captures the gener-

---

[1] https://github.com/wilmsm/lightweightbraindiff

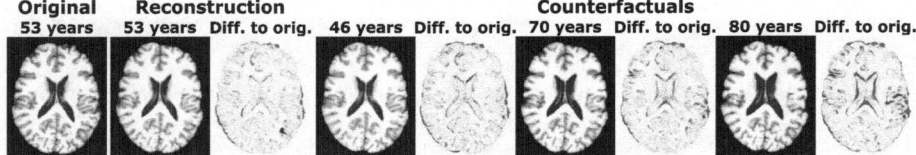

Fig. 2. Brain age prediction explainability results for one test subject (53 yrs.) when using the adult DWPT-enhanced conditional diffusion model. The model correctly predicts the age of the original image while not being able to fully represent all of its details when reconstructed by the model (reconstruction = 'counterfactual' at the predicted age). The counterfactual images and the associated difference images to the original visually highlight what the model would have expected to see for a prediction at those ages (46/70/80 yrs. instead of 53 yrs.). Difference images: blue – negative gray value diff., red – positive diff., white – no diff.) (Color figure online)

ally expected trend of healthy aging and its variability (*e.g.*, expanding lateral ventricles and sulci widening with increasing age). This is also confirmed by the quantitative results for the brain age prediction task. Here, the adult diffusion model achieves a MAE of 3.93 ± 3.00 years while the pediatric model's MAE is 1.38 ± 1.24 years. In the presented setup with 50 Monte Carlo samples, age prediction takes 16 s/image for the adult model and 8 s/image for the pediatric model (incl. DWPT, which only adds a negligible overhead < 1 s; SFCN: < 1 s). Predictions with the adult model require more time due to the larger number of age classes. Both MAEs are slightly worse than those achievable with the purely discriminative SFCN models (adult MAE: 2.79 ± 2.10 yrs., pediatric MAE: 1.07 ± 0.89 yrs.), which is to be expected based upon previously reported results with generative brain age prediction models (*e.g.*, [29,30]). However, the numbers achieved by our model are to our best knowledge the best results that have ever been reported for generative brain age prediction models in adults (*e.g.*, 4.83 yrs. in [30] and 4.50 yrs. in [29]) and we are the first to report results for pediatric brain age prediction with generative models. Moreover, our generative models are inherently explainable through their ability to generate counterfactual images that visually explain their decisions. This is exemplarily showcased in Fig. 2 for one individual from the adult test set. The visualizations do not only confirm that the proposed counterfactual mechanism preserves the general identity features of the brain when changing the age value, but also highlight again the faithfulness of the aging mechanism learned by the model (*e.g.*, enlargement of sulci and ventricles with increasing age).

4   Conclusion

In this paper, we present a lightweight wavelet-based conditional diffusion model architecture that enables computationally efficient training of those models on full-resolution 3D T1-weighted brain MRI data on a single consumer-grade GPU. The key technical idea of the proposed lightweight architecture is to substantially reduce the spatial size of the MR images that are processed by the diffusion model's denoiser without losing any information. This is achieved through a fixed wavelet packet transform and our proof-of-concept evaluation validates this idea. Moreover, we are the first to explore the use of such conditional diffusion models for self-explainable brain age prediction in adults and children. Our results show that the models achieve better accuracy than what has been previously reported for brain age prediction with generative models and are able produce realistic counterfactual images for highly accessible self-explainability. Given the proof-of-concept character of our work, future work will focus, for example, on additional (quantitative) comparisons to other generative and discriminative approaches in this area as well as the exploration of additional strategies to improve the models' accuracy and computational efficiency – especially during age prediction. This could include the exploration of smaller, even more optimized network architectures.

Acknowledgements. This work was supported in part by the Azrieli Accelerator at the University of Calgary, which was established through a generous gift from the Azrieli Foundation, the Department of Pediatrics at the University of Calgary, the Alberta Children's Hospital Foundation, and the Natural Sciences and Engineering Research Council of Canada (NSERC). This research has been conducted using the UK Biobank resource under application number 77508. Data used in the preparation of this article were obtained from the Adolescent Brain Cognitive Development (ABCD) Study (https://abcdstudy.org), held in the NIMH Data Archive (NDA). This is a multisite, longitudinal study designed to recruit more than 10,000 children age 9–10 and follow them over 10 years into early adulthood. The ABCD Study is supported by the National Institutes of Health and additional federal partners under award numbers U01DA041048, U01DA050989, U01DA051016, U01DA041022, U01DA051018, U01DA-051037, U01DA050987, U01DA041174, U01DA041106, U01DA041117, U01DA041028, U01DA041134, U01DA050988, U01DA051039, U01DA041156, U01DA041025, U01DA-041120, U01DA051038, U01DA041148, U01DA041093, U01DA041089, U24DA041123, U24DA041147. A full list of supporters is available at https://abcdstudy.org/federal-partners.html. A listing of participating sites and a complete listing of the study investigators can be found at https://abcdstudy.org/consortium_members/. ABCD consortium investigators designed and implemented the study and/or provided data but did not necessarily participate in the analysis or writing of this report. This manuscript reflects the views of the authors and may not reflect the opinions or views of the NIH or ABCD consortium investigators.

Disclosure of Interests. The authors have no competing interests to declare.

References

1. Alexander, L.M., et al.: An open resource for transdiagnostic research in pediatric mental health and learning disorders. Sci. Data **4**(1), 1–26 (2017)
2. Avants, B.B., et al.: The optimal template effect in hippocampus studies of diseased populations. Neuroimage **49**(3), 2457–2466 (2010)
3. Bannister, J.J., et al.: A deep invertible 3-D facial shape model for interpretable genetic syndrome diagnosis. IEEE JBHI **26**(7), 3229 (2022)

4. Bhattacharya, M., Prasanna, P.: GazeDiff: a radiologist visual attention guided diffusion model for zero-shot disease classification. In: MIDL (2024)
5. Casey, B.J., et al.: The adolescent brain cognitive development (ABCD) study: imaging acquisition across 21 sites. Dev. Cogn. Neurosci. **32**, 43–54 (2018)
6. Clark, K., Jaini, P.: Text-to-image diffusion models are zero shot classifiers. NeurIPS **36** (2023)
7. Di Martino, A., et al.: Enhancing studies of the connectome in autism using the autism brain imaging data exchange II. Sci Data **4**(1), 1–15 (2017)
8. Dorjsembe, Z., Pao, H.K., Odonchimed, S., Xiao, F.: Conditional diffusion models for semantic 3D medical image synthesis. arXiv:2305.18453 (2023)
9. Fonov, V.S., Evans, A.C., McKinstry, R.C., Almli, C.R., Collins, D.L.: Unbiased nonlinear average age-appropriate brain templates from birth to adulthood. Neuroimage **47**, S102 (2009)
10. Friedrich, P., Wolleb, J., Bieder, F., Durrer, A., Cattin, P.C.: WDM: 3D wavelet diffusion models for high-resolution medical image synthesis. arXiv:2402.19043 (2024)
11. Ho, J., Jain, A., Abbeel, P.: Denoising diffusion probabilistic models. NeurIPS **33**, 6840–6851 (2020)
12. Isensee, F., et al.: Automated brain extraction of multisequence MRI using artificial neural networks. Hum. Brain Mapp. **40**(17), 4952–4964 (2019)
13. Jaini, P., Clark, K., Geirhos, R.: Intriguing properties of generative classifiers. arXiv:2309.16779 (2023)
14. Jeyakumar, J.V., Noor, J., Cheng, Y.H., Garcia, L., Srivastava, M.: How can I explain this to you? An empirical study of deep neural network explanation methods. NeurIPS **33** (2020)
15. Kazerouni, A., et al.: Diffusion models in medical imaging: a comprehensive survey. Med. Image Anal., 102846 (2023)
16. Khader, F.: Denoising diffusion probabilistic models for 3D medical image generation. Sci. Rep. **13**(1), 7303 (2023)
17. Li, A.C., Prabhudesai, M., Duggal, S., Brown, E., Pathak, D.: Your diffusion model is secretly a zero-shot classifier. In: ICCV 2023, pp. 2206–2217 (2023)
18. Li, Q., Shen, L., Guo, S., Lai, Z.: Wavelet integrated CNNs for noise-robust image classification. In: CVPR 2020, pp. 7245–7254 (2020)
19. Littlejohns, T.J., et al.: The UK biobank imaging enhancement of 100,000 participants: rationale, data collection, management and future directions. Nature Comms **11**(1), 2624 (2020)
20. Mallat, S.: A Wavelet Tour of Signal Processing. Elsevier, Amsterdam (1999)
21. Peng, H., Gong, W., Beckmann, C.F., Vedaldi, A., Smith, S.M.: Accurate brain age prediction with lightweight deep neural networks. Med. Image Anal. **68**, 101871 (2021)
22. Phung, H., Dao, Q., Tran, A.: Wavelet diffusion models are fast and scalable image generators. In: CVPR 2023, pp. 10199–10208 (2023)
23. Pinaya, W.H., et al.: Generative AI for medical imaging: extending the MONAI framework. arXiv:2307.15208 (2023)
24. Pinaya, W.H.L., et al.: Brain imaging generation with latent diffusion models. In: Mukhopadhyay, A., Oksuz, I., Engelhardt, S., Zhu, D., Yuan, Y. (eds.) DGM4MICCAI 2022. LNCS, vol. 13609, pp. 117–126. Springer, Cham (2022)
25. Rudin, C.: Stop explaining black box machine learning models for high stakes decisions and use interpretable models instead. Nat. Mach. Intell. **1**(5), 206–215 (2019)

26. Singla, S., Eslami, M., Pollack, B., Wallace, S., Batmanghelich, K.: Explaining the black-box smoothly-a counterfactual approach. Med. Image Anal. **84**, 102721 (2023)
27. Song, J., Meng, C., Ermon, S.: Denoising diffusion implicit models. arXiv:2010.02502 (2020)
28. Van der Velden, B.H., Kuijf, H.J., Gilhuijs, K.G., Viergever, M.A.: Explainable artificial intelligence (XAI) in deep learning-based medical image analysis. Med. Image Anal. **79**, 102470 (2022)
29. Wilms, M., et al.: Invertible modeling of bidirectional relationships in neuroimaging with normalizing flows: application to brain aging. IEEE Trans. Med. Imaging **41**(9), 2331–2347 (2022)
30. Wilms, M., Mouches, P., Bannister, J.J., Rajashekar, D., Langner, S., Forkert, N.D.: Towards self-explainable classifiers and regressors in neuroimaging with normalizing flows. In: International Workshop on Machine Learning in Clinical Neuroimaging, pp. 23–33 (2021)
31. Wu, W., Wang, Y., Liu, Q., Wang, G., Zhang, J.: Wavelet-improved score-based generative model for medical imaging. IEEE Trans. Med. Imaging **43**, 966–979 (2023)
32. Zhou, S.K., et al.: A review of deep learning in medical imaging: imaging traits, technology trends, case studies with progress highlights, and future promises. Proc. IEEE **109**(5), 820–838 (2021)

SOE: SO(3)-Equivariant 3D MRI Encoding

Shizhe He[1], Magdalini Paschali[2], Jiahong Ouyang[3], Adnan Masood[4], Akshay Chaudhari[2,5], and Ehsan Adeli[1,3(✉)]

[1] Department of Computer Science, Stanford University, Stanford, CA, USA
eadeli@stanford.edu
[2] Department of Radiology, Stanford University, Stanford, CA, USA
[3] Department of Psychiatry and Behavioral Sciences, Stanford University, Stanford, CA, USA
[4] UST, Aliso Viejo, CA, USA
[5] Department of Biomedical Data Science, Stanford University, Stanford, CA, USA

Abstract. Representation learning has become increasingly important, especially as powerful models have shifted towards learning latent representations before fine-tuning for downstream tasks. This approach is particularly valuable in leveraging the structural information within brain anatomy. However, a common limitation of recent models developed for MRIs is their tendency to ignore or remove geometric information, such as translation and rotation, thereby creating invariance with respect to geometric operations. We contend that incorporating knowledge about these geometric transformations into the model can significantly enhance its ability to learn more detailed anatomical information within brain structures. As a result, we propose a novel method for encoding 3D MRIs that enforces equivariance with respect to all rotations in 3D space, in other words, SO(3)-equivariance (SOE). By explicitly modeling this geometric equivariance in the representation space, we ensure that any rotational operation applied to the input image space is also reflected in the embedding representation space. This approach requires moving beyond traditional representation learning methods, as we need a representation vector space that allows for the application of the same SO(3) operation in that space. To facilitate this, we leverage the concept of vector neurons. The representation space formed by our method, SOE, captures the brain's structural and anatomical information more effectively. We evaluate SOE pretrained on the structural MRIs of two public data sets with respect to the downstream task of predicting age and diagnosing Alzheimer's Disease from T1-weighted brain scans of the ADNI data set. We demonstrate that our approach not only outperforms other methods but is also robust against various degrees of rotation along different axes. The code is available at https://github.com/shizhehe/SOE-representation-learning.

Keywords: Contrastive Learning · SO(3)-Equivariance · 3D Brain MRI

© The Author(s), under exclusive license to Springer Nature Switzerland AG 2025
D. R. Bathula et al. (Eds.): MLCN 2024, LNCS 15266, pp. 68–77, 2025.
https://doi.org/10.1007/978-3-031-78761-4_7

1   Introduction

In recent years, there has been a significant surge in the development and adoption of various self-supervised representation learning techniques [2,7,9,17,21]. These methods aim to create universal approaches (a.k.a. foundation models) for extracting meaningful features or representations from unlabeled data. This approach has gained popularity due to its ability to utilize large amounts of readily available unlabeled data, reducing the dependency on costly and time-consuming labeling processes. Various pretraining techniques have recently also been developed for medical images [6,10,18]. The recent methods developed for medical images, including magnetic resonance imaging (MRI), have mostly replicated pretraining techniques used in natural images applications. Examples of such are MRI masked autoencoders [14].

A prevalent issue with recent models designed for MRI data is their tendency to overlook or eliminate geometric information, such as translation and rotation (e.g., [5,13]), resulting in invariance to geometric transformations. We argue that making MRI encoders understand these geometric transformations can improve their capacity to capture intricate anatomical details within brain structures.

In this paper, we introduce a novel approach for encoding 3D MRIs that enforces SO(3)-equivariance (SOE).

SO(3) refers to the special orthogonal group in three dimensions, which represents all possible rotations in 3D space [6]. Mathematically, it consists of all 3×3 orthogonal matrices with a determinant of $+1$. Traditionally, in computer vision, graphics, robotics, and physics, these matrices describe the rotation of an object in 3D space without any reflection or scaling. SO(3) rotations are commonly used to represent the orientation of objects or coordinate systems [3,6,22]. By explicitly incorporating geometric equivariance, we ensure that any operation performed on the input image space is correspondingly mirrored in the embedding representation space. This strategy necessitates a departure from conventional representation learning methods, as it requires a representation vector space (or vector neurons) that accommodates the application of the same SO(3) operation in that space.

In summary, our contributions are: (1) We propose a method to learn geometry-aware representations by modeling rotation through SO(3) matrix transformation and enforcing SO(3)-equivariance between the original 3D MRI space with the latent representation space (see Fig. 1). Here, we introduce a technique to regularize the model to prevent it from falling into trivial solutions that completely ignore SO(3) operations. (2) We formulate this self-supervised learning framework within a model architecture composed of a generic encoder, the Vector Neuron module [3], and a prediction head. (3) We show that incorporating geometric information during pretraining improves the performance on AD classification and age prediction tasks.

2 Methods

Here, we present SO(3)-Equivariant (SOE) Representation Learning for 3D MRI encoding. Let $Rot(\cdot)$ be a 3D rotation function that maps a 3D medical scan to its rotated orientation and x_i^1 be a 3D MRI input whose representation is generated using the encoder network f_θ. Given an arbitrary rotated version of x_i^1 by $x_i^2 = Rot(x_i^1)$, we argue that optimizing for rotational equivariance between the representations $f_\theta(x_i^1) = z_i^1$ and $f_\theta(x_i^2) = z_i^2$, the encoder learns structural coherence in the 3D MRI. To enforce this equivariance, we create a self-supervised setup, in which the same $Rot(\cdot)$ transformation applies to both spaces, i.e., the MRI input and the latent representation. To enable this, we need to turn the latent space into a vector space to enable the application of matrix transformation operations. To this end, we supplement our generic encoder with a Vector Neuron (VN) module [3].

First, we describe our approach to modeling 3D image rotation $Rot(\cdot)$ as matrix transformations of the SO(3) group in Sect. 2.1. We then define our proposed pretext representation learning objective and offer a more detailed insight into the VN module [3].

Next, we propose an inverse maximization regularizer for the pretext training objective to prevent the representationsfrom converging to the same point in feature space and avoid trivial solutions. Finally, we introduce a robustness regularization term for downstream task training.

2.1 Modeling 3D Rotation as SO(3) Matrix Transformations

Let R_{rot} be a 3×3 rotation matrix defined by an axis and rotation angle. There exists no straight-forward method (such as matrix multiplication) of applying R_{rot} on a three-dimensional image volume x^1 of shape $n \times n \times n$ in non-coordinate representation, where n is the dimensionality of the 3D volume. We define $Rot(\cdot)$ (Fig. 1) as follows: (1) mapping x^1 to its coordinate representation y^1 of shape $n \times n \times n \times 3$ where the entries $y_{(i,j,k,c)}^1$ are centered around the origin of the coordinate system, (2) applying R_{rot} to the spatial coordinate representation of each pixel $y_{(i,j,k,c)}^1$ and thereby computing the new rotated location (p_1, p_2, p_3) of each voxel within the image grid $y_{(i,j,k,c)}^2$, (3) mapping and clamping the spatial coordinate representation y^2 into the image grid x^2, and (4) performing trilinear interpolation for pixel coordinates (p_1, p_2, p_3) that do not fit into exact entries of the image grid. Since R_{rot} is defined by the axis and angle of the applied rotation, we omit them from the definition of $Rot(\cdot)$ for simplicity. Therefore, the rotated volume x^2 is defined as:

$$x_{(i,j,k,c)}^2 = Rot(x_{(i,j,k,c)}^1, R_{rot}) = \text{Interpolation}\left(R_{rot}x_{(i,j,k,c)}^1\right). \tag{1}$$

2.2 SO(3)-Equivariant Representation Learning

We define a pretext training objective to learn representations that optimize towards SO(3)-equivariance of the 3D MRIs and their feature representations. To

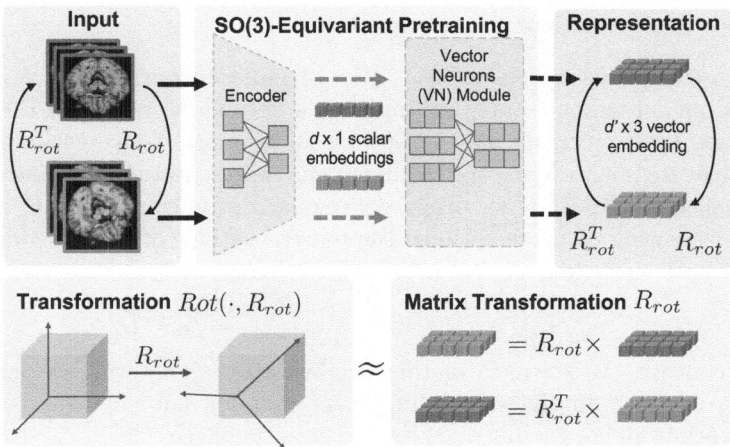

Fig. 1. SOE Model Architecture: The model takes in a pair of unrotated and rotated samples x^1, x^2 as input and maps them to representations preserving the rotational transformation R_{rot}. The Encoder encodes the input volume into a d dimensional scalar array and the VN module maps that scalar array into a $d' \times 3$ dimensional vector embedding. R_{rot} is applied on the input volume through the $Rot(\cdot)$ function.

do so, we define our feature encoder (Fig. 1) $f(\cdot)$ to map the input $n_{vol} \times n_{vol} \times n_{vol}$ 3D volume to the corresponding $n_{hidden} \times 1$ dimensional representation. Furthermore, we define our Vector Neuron module (VN), which maps the $n_{hidden} \times 1$ dimensional scalar representation to a $n_{hidden} \times 3$ vector representation (details in Sect. 2.3). With x^2 being the rotated version of x^1, we define the self-supervised representation learning objective to be

$$\mathcal{L}_{SO(3)} = \|\text{VN}(f(x^1))R_{rot} - \text{VN}(f(x^2))\|^2 + \|\text{VN}(f(x^1)) - \text{VN}(f(x^2))R_{rot}^T\|^2. \quad (2)$$

This loss function enforces rotational equivariance in the representation space by minimizing the Frobenius distance between $VN(f(x^2))$, the representation of the rotated input sample x^2, and $VN(f(x^1))R_{rot}$, the representation of the unrotated input sample transformed to approximate the representation of x^2. We model the representations to preserve the symmetry of SO(3)-matrix/rotational transformations from image to feature space.

2.3   Vector Neuron Module

The Vector Neuron (VN) module (see Fig. 1, middle) plays a pivotal role in our self-supervised learning task as described in Sect. 2.2. We use VN [3] to map $d \times 1$ scalar representations to $d' \times 3$ vector representations of the input samples where d is the output dimension of the encoder. In other words, the VN layers project the conventional scalar neurons into 3D vector neurons to enable enforcing full SO(3)-equivariance between the input and the representation space. This way, we can use vector representation as 3D points to compute our self-supervised equivariance loss $\mathcal{L}_{SO(3)}$ (Eq. (2)).

2.4   Inverse Maximization Component

The proposed equivariance loss $\mathcal{L}_{SO(3)}$ (Eq. (2)) may encourage feature collapse, where the model learns to map all input to a very small region in the feature space, since $\mathcal{L}_{SO(3)}$ can be naively optimized as inputs converge to the same point in feature space. This would hinder our representation learning process. To avoid such trivial cases, we propose a regularization component to inversely penalize decreasing distance between the representations of x^1 and x^2:

$$\mathcal{L}_{comb}(x^1, x^2) = \mathcal{L}_{SO(3)} + \lambda \frac{1}{\|f(x^1) - f(x^2)\|^2}, \tag{3}$$

where λ controls the strength of this regularization. This way, we intuitively encourage diversity in the embedding space and prevent all instances in the feature space from converging to the same point in space.

2.5   Robustness Regularizer Against Rotations

One of the main benefits of SOE is the robustness against 3D rotations. Here, we define the notion of robustness regularization during the downstream training of our model. We need to explicitly incorporate this aspect beyond merely modeling SO(3)-equivariance in the representation space during pretext training since traditional downstream loss functions do not learn/preserve robustness against geometric transformations by themselves [11]. Rotational robustness is when a trained model is able to consistently predict the correct label/age under all degrees of rotations. We define robustness in this context as directly incorporating robustness modeled as SO(3)-invariance into the downstream objective function as a regularizer. In other words, the terms introduced in Eq. (2) will be added to the downstream fine-tuning objective.

3   Experimental Results

We design various experiments to examine the performance and generalizability of our representation learning approach.

Datasets. We evaluate our approach using two 3D brain MRI datasets for pretraining with representation learning and one dataset for pretraining and subsequent fine-tuning on two downstream tasks. The Alzheimer's Disease Neuroimaging Initiative (ADNI) dataset [16] consists of 2,577 T1-weighted MRIs of 811 subjects, where each subject has two to six visits. The first downstream task on ADNI is disease classification into four classes: cognitively normal (NC) (N = 214), Alzheimer's disease (AD) (N = 187), static MCI (sMCI) (N = 275), and progressive MCI (pMCI) (N = 135). The subject age ranges between 54.4 and 90.9 (mean 75.18 ± 6.84). The second downstream task on ADNI is age regression.

The NCANDA study [20] recruited 831 participants across five sites in the United States and performed yearly imaging assessments [1]. In this work, we used 3,830 T1-weighted MRIs of 831 subjects. The NCANDA dataset was used solely for pretext representation learning. We downsample both datasets to a resolution of 64×64×64 [18]. We split the datasets into training (70%), validation (10%), and testing (20%) sets and into five folds for stratified cross-validation by subject [18].

Implementation. All self-supervised pretraining approaches are based on a convolutional encoder backbone composed of four convolution blocks, amounting to 97,584 trainable parameters. Each of the convolution blocks consists of 3D Conv (kernel size $3 \times 3 \times 3$), 3D BatchNorm, LeakyReLU/ReLU (slope 0.2), 3D Dropout, and 3D MaxPool (kernel size 2) layers built on PyTorch version 2.0.1 [19]. All models were trained on an NVIDIA GeForce RTX 2080 Ti GPU. Note that we also examined newer transformer-based models, namely the 3D SWIN Transformer [15] and Vision Transformer [4] as the backbone encoder in SOE. However, the results show no significant increase against the convolutional encoder backbone. Hence, for the sake of simplicity and consistency, we make all comparisons using conventional backbones.

During pretraining for both SOE and baseline models, the mini-batch size is set to either 64 or 32. For the downstream tasks, the mini-batch size is set to 64, All models are trained for a maximum of 50 epochs for both pretext and downstream training. The learning rates are initialized to a value varying between 0.01 and 0.0001, and decrease in logarithmic steps during training.

Baselines. To examine the performance of our proposed approach, we compare our model against other representation learning approaches based on the downstream metrics. For the classification tasks we report balanced accuracy (BACC), F1 score and their standard deviation across folds. For regression we measure coefficient of determination (R2), and mean absolute error (MAE).

We use the following baseline models as a comparison to SOE: (1) Autoencoder (AE), which learns efficient low-dimensional representations; (2) Variational Autoencoder (VAE), which learns a continuous latent space for its input data [12]; (3) Masked Autoencoder (MAE), a variant of AE where random patches of the input are masked during training [8]; and (4) SimCLR, a contrastive self-supervised learning framework [2].

Robustness Experiments. We perform rotational robustness experiments comparing SOE to the baseline model with no pretraining on different levels of rotation. Both models are trained with the robustness regularizer to incorporate rotational robustness in downstream task optimization. We evaluate the models' performance against (1) mild rotations chosen randomly between 15 and 45° and (2) rotations in 90° intervals from 0. Note that rotations not in 90° intervals from 0 will experience interpolation as described in Sect. 2.1. We train the robustness models on the two levels of rotation and test them on no rotation and the respective rotations.

3.1 Pretraining and Dataset Impact

First, we evaluated our proposed method with regard to its generalizability across datasets. We examined the classification and age regression performance

Table 1. NCANDA vs. ADNI pretraining, downstream task evaluated on ADNI. (a) NC vs. AD classification cross-validated; (b) NC age regression cross-validated. Our model (SOE) outperforms all other models across both tasks.

(a) Classification			(b) Age Regression		
Pretraining	BACC$^\uparrow$	F1$^\uparrow$	Pretraining	R2$^\uparrow$	MAE (years)$^\downarrow$
-	82.34 ± 4.24	78.15 ± 5.56	-	0.36 ± 0.16	6.37 ± 0.04
NCANDA	82.81 ± 2.01	78.50 ± 2.92	NCANDA	0.41 ± 0.09	5.95 ± 0.06
ADNI	**83.84 ± 4.11**	**79.38 ± 5.58**	ADNI	**0.45 ± 0.03**	**5.88 ± 0.04**

on ADNI of models without pretraining, pretrained with SOE on the NCANDA dataset, and pretrained with SOE on the ADNI dataset. In Table 1, we observe that our proposed SOE pretraining performs better than the no pretraining instance for both ADNI NC vs. AD classification and age prediction tasks by 0.5–8% in terms of BACC. Furthermore, we observe that pretraining our model on the same dataset (ADNI) as the downstream task evaluation, the so-called "self-training", outperforms the model pretrained on NCANDA in classification (AUC 91.3 vs. 91.1; BACC 83.84 vs. 82.81; F1 79.38 vs. 78.50). Particularly when evaluated on the task of age prediction, our representation learning framework outperforms the model with no pretraining (R2 0.45 vs. 0.41; MAE 5.88 vs. 5.95). Consequently, this confirms our expectation that SOE representation learning allows the model to extract useful information and map to expressive representations. We decide to continue with the "self-training" setup, i.e., self-supervisd pretraining and downstream training on ADNI, for all further experiments.

3.2   Classification Results on ADNI

After pretraining on the ADNI dataset, we evaluated the proposed SOE framework against the baseline models on ADNI NC vs. AD binary classification as a downstream task. As shown in Table 2, we observe that our best SOE model with robustness regularizer (Sect. 2.5) outperforms all other compared models by 1.1–4.7% in BACC, 1.9–3.7% in AUC, and 1.3–6.2% in F1 Score. Specifically, the best SOE model without the robustness regularizer performs on par with SimCLR model and marginally worse than the best MAE model. This aligns with our hypothesis that explicitly modeling the SO(3)-equivariance in the representation space is key to leveraging valuable geometric information of 3D MRIs compared to approaches that neglect geometry altogether.

3.3   Age Regression Results on ADNI

Next, we evaluated the proposed representation learning framework with regards to the age prediction task on the ADNI NC cohort. Under the NC age regression column in Table 2, we observe that this regression task is quite challenging overall. We observe marginal improvements in MAE and R2 coefficient of our

Table 2. ADNI classification and age regression comparison with SOTA methods.

	NC vs. AD Classification		NC Age Regression	
	BACC$^\uparrow$	F1$^\uparrow$	R2$^\uparrow$	MAE$^\downarrow$
No pretraining	82.34 ± 4.24	78.15 ± 5.56	0.36 ± 0.16	6.37 ± 0.04
AE	81.63 ± 4.62	76.68 ± 6.37	**0.46 ± 0.02**	5.69 ± 0.00
VAE	83.14 ± 3.81	79.06 ± 5.03	0.45 ± 0.02	5.70 ± 0.00
MAE	84.69 ± 2.28	80.75 ± 3.69	**0.46 ± 0.05**	**5.67 ± 0.02**
SimCLR	83.94 ± 4.03	79.73 ± 4.47	0.42 ± 0.09	6.07 ± 0.11
SOE w/o reg.	83.84 ± 4.11	79.38 ± 5.58	0.45 ± 0.03	5.88 ± 0.04
Proposed Method (SOE)	**85.62 ± 2.80**	**81.78 ± 4.21**	0.45 ± 0.03	5.88 ± 0.04

best SOE model against the baseline pretrained models VAE and SimCLR. AE
and MAE models marginally outperform our proposed model by less than 3%
in terms of MAE and R2.

3.4 Robustness Against Rotation

We further examined the robustness of the proposed SOE representation learning framework at different levels of rotations as data augmentation during downstream training. Specifically, we compare our best model against a model with
no pretraining when trained on the ADNI classification task with different levels
of rotation augmentation and evaluated on different levels of rotation. As seen in
Fig. 2, we observe increase in BACC and F1 for the SOE pretrained model when
trained with rotation augmentations and evaluated on unrotated samples. Similarly, we see that the proposed SOE representation learning framework displays

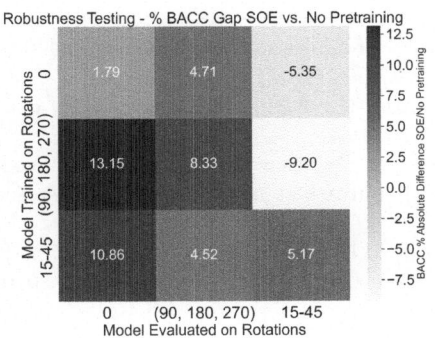
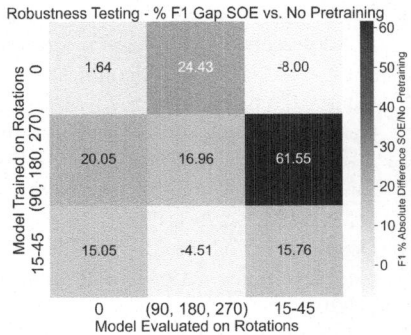

Fig. 2. ADNI Classification: Comparison of rotation augmented SOE vs. no pretraining
model. Displayed are the % increases in BACC (left) and F1 (right) from a model
with no pretraining to a SOE pretrained model. The labels on each x axis shows the
rotation degrees employed during training and the y axis rotation degrees employed
during evaluation.

superior performance over a no pretraining approach in nearly all other scenarios. This confirms our expectation that when augmented with rotations during downstream training, a SO(3)-equivariant encoder can more easily adapt to rotations and implicitly learn rotational robustness than an untrained encoder. In theory, the classification head on a SO(3)-equivariant encoder would only have to learn SO(3)-invariance to ensure rotational robustness.

4   Conclusion

We proposed a self-supervised representation learning framework, SOE, which models geometric SO(3)-equivariance in the feature representation space by leveraging the Vector Neuron Framework. By modeling SO(3)-equivariance, we move beyond conventional representation learning methods which neglect geometric information. Experiments show the generalizability of our representation learning approach across datasets and showed that SOE outperforms other SOTA representation learning methods on AD classification and is on par on age regression. Additionally, we illustrated the potential of SOE to facilitate more extreme rotations for data augmentation and its potential to increase model robustness against different levels of rotation. A direction for future work is to expand the notion of equivariance to more distortions that can be modeled as matrix transformations to enforce further geometric and mathematical coherence between the input and latent spaces.

Acknowledgements. This work was supported by the HAI-Google Research Award, U.S. National Institute (AG073104), the 2024 HAI Hoffman-Yee Grant, and the HAI-Google Cloud Credits Award.

References

1. Brown, S.A., et al.: The national consortium on alcohol and neurodevelopment in adolescence (NCANDA): a multisite study of adolescent development and substance use. J. Stud. Alcohol Drugs **76**(6), 895–908 (2015)
2. Chen, T., Kornblith, S., Norouzi, M., Hinton, G.: A simple framework for contrastive learning of visual representations (2020)
3. Deng, C., Litany, O., Duan, Y., Poulenard, A., Tagliasacchi, A., Guibas, L.: Vector neurons: a general framework for so(3)-equivariant networks. arXiv preprint arXiv:2104.12229 (2021)
4. Dosovitskiy, A., et al.: An image is worth 16x16 words: transformers for image recognition at scale. CoRR abs/2010.11929 (2020). https://arxiv.org/abs/2010.11929
5. Esfahani, E.E., Hosseini, A.: Compressed MRI reconstruction exploiting a rotation-invariant total variation discretization. Magn. Reson. Imaging **71**, 80–92 (2020)
6. Esteves, C., Allen-Blanchette, C., Makadia, A., Daniilidis, K.: Learning so(3) equivariant representations with spherical CNNs. In: Proceedings of the European Conference on Computer Vision (ECCV) (2018)
7. Grill, J.B., et al.: Bootstrap your own latent: a new approach to self-supervised learning (2020)

8. He, K., Chen, X., Xie, S., Li, Y., Dollár, P., Girshick, R.: Masked autoencoders are scalable vision learners (2021)
9. He, K., Fan, H., Wu, Y., Xie, S., Girshick, R.B.: Momentum contrast for unsupervised visual representation learning. CoRR abs/1911.05722 (2019). http://arxiv.org/abs/1911.05722
10. Huang, S.C., Pareek, A., Jensen, M., Lungren, M.P., Yeung, S., Chaudhari, A.S.: Self-supervised learning for medical image classification: a systematic review and implementation guidelines. NPJ Digit. Med. **6**(1), 74 (2023)
11. Ilyas, A., Santurkar, S., Tsipras, D., Engstrom, L., Tran, B., Madry, A.: Adversarial examples are not bugs, they are features (2019)
12. Kingma, D.P., Welling, M.: Auto-encoding variational Bayes (2022)
13. Kwon, S., Choi, J.Y., Ryu, E.K.: Rotation and translation invariant representation learning with implicit neural representations. arXiv preprint arXiv:2304.13995 (2023)
14. Lang, D.M., Schwartz, E., Bercea, C.I., Giryes, R., Schnabel, J.A.: 3D masked autoencoders with application to anomaly detection in non-contrast enhanced breast MRI. arXiv preprint arXiv:2303.05861 (2023)
15. Liu, Z., et al.: Swin transformer: Hierarchical vision transformer using shifted windows. CoRR abs/2103.14030 (2021). https://arxiv.org/abs/2103.14030
16. Mueller, S.G., et al.: The Alzheimer's disease neuroimaging initiative. Neuroimaging Clin. N. Am. **15**(4), 869–877 (2005). https://doi.org/10.1016/j.nic.2005.09.008, https://www.sciencedirect.com/science/article/pii/S1052514905001024, alzheimer's Disease: 100 Years of Progress
17. van den Oord, A., Li, Y., Vinyals, O.: Representation learning with contrastive predictive coding (2019)
18. Ouyang, J., Zhao, Q., Adeli, E., Zaharchuk, G., Pohl, K.M.: Self-supervised learning of neighborhood embedding for longitudinal mri. Med. Image Anal. **82**, 102571 (2022). https://doi.org/10.1016/j.media.2022.102571, https://www.sciencedirect.com/science/article/pii/S1361841522002122
19. Paszke, A., et al.: Pytorch: an imperative style, high-performance deep learning library. In: Advances in Neural Information Processing Systems, vol. 32, pp. 8024–8035. Curran Associates, Inc. (2019). http://papers.neurips.cc/paper/9015-pytorch-an-imperative-style-high-performance-deep-learning-library.pdf
20. Pohl, K.M., et al.: The NCANDA_PUBLIC_6Y_REDCAP_V01 data release of the national consortium on alcohol and neurodevelopment in adolescence (NCANDA) (2021). https://dx.doi.org/10.7303/syn25606546
21. Sriram, A., Gaidon, A., Wu, J., Niebles, J.C., Fei-Fei, L., Adeli, E.: Home: homography-equivariant video representation learning (2023)
22. Weiler, M., Hamprecht, F.A., Storath, M.: Learning steerable filters for rotation equivariant CNNs. In: 2018 IEEE/CVF Conference on Computer Vision and Pattern Recognition, pp. 849–858 (2018). https://doi.org/10.1109/CVPR.2018.00095

Towards a Foundation Model for Cortical Folding

Julien Laval[1]($\boxtimes$), Joël Chavas[1], Vanessa Troiani[2], William Snyder[3,4],
Marisa Patti[5], Mylène Moyal[6], Marion Plaze[6], Arnaud Cachia[7],
Zhong Yi Sun[8], Vincent Frouin[1], Pietro Gori[9], Denis Rivière[1],
and Jean-François Mangin[1]

[1] Université Paris-Saclay, CEA, CNRS, NeuroSpin, U9027 Baobab, Saclay, France
laval.julien1@gmail.com
[2] Geisinger Autism and Developmental Medicine Institute, Lewisburg, USA
[3] National Institute of Mental Health Intramural Research Program, Bethesda, USA
[4] Department of Psychiatry, University of Cambridge, Cambridge, UK
[5] A.J. Drexel Autism Institute, Drexel University, Philadelphia, USA
[6] GHU Paris, IPNP, INSERM U1266, Paris, France
[7] Université de Paris, LaPsyDÉ, CNRS, Paris, France
[8] US52-UAR2031 CATI, Institut du Cerveau (ICM), Paris, France
[9] LTCI, Télécom Paris, Institut Polytechnique de Paris, Palaiseau, France

Abstract. The brain surface is composed of humps called gyri, separated by grooves called sulci. Although the main folds are common to all individuals, their shape varies, making them unique to each individual. Cortical folding may contain biomarkers that have yet to be deciphered. While conventional geometric approaches fail to fully characterize the high inter-individual variability, recent efforts in large-scale MRI data collection allow us to leverage the statistical power of deep neural networks. Here, we introduce Champollion V0, a self-supervised learning (SSL) algorithm to sort sulcal variability based on 21,070 subjects from the UKBioBank dataset. We revisit from scratch an existing model and optimize its ability to retrieve hand-labeled patterns defined by the neuroscientific community. Under linear evaluation on the latent space, Champollion V0 significantly improves the detection of three different kinds of folding patterns: the presence of a parallel sulcus (AUC increases from 73% to 84%), the presence of specific interruptions (AUC increases from 50% to 79%) and the detection of a specific folding shape (R^2 increases on each of the six main geometric features), respectively in the cingulate, the orbital and the central region. These hand-labeled patterns were found to be correlated to neurodevelopmental pathologies. Champollion V0 could enable the automatic labeling of larger datasets for future studies. The code can be found on Github.

Keywords: Brain · folding patterns · MRI · Self-supervised learning

Supplementary Information The online version contains supplementary material available at https://doi.org/10.1007/978-3-031-78761-4_8.

1   Introduction

The brain is folded, formed of bumps (gyri) separated by folds (sulci). The major folds are common to all individuals. However, they vary in shape, making each brain unique. Some sulcal patterns have been identified as biomarkers: for example, the cingulate region can display either a single main sulcus, or an additional paracingulate sulcus (PCS), parallel to the cingulate (Fig. 1a). The asymmetry of this pattern between the two hemispheres has been linked to control efficiency in preschoolers [4]. Similarly, different prevalences of the orbitofrontal cortex (OFC) interruption types (Fig. 1b) have been reported in schizophrenic and catatonic subjects compared to controls [16,18].

The brain folds may contain many biologically relevant patterns that have yet to be discovered. However, identifying and labeling such patterns is very time-consuming, if not impossible for human experts. The recent acquisitions of large MRI datasets have enabled the use of deep learning to sort the sulcal variability and extract patterns automatically. Gaudin et al. developed a contrastive self-supervised (SSL) model (named thereafter the Orig. model) to encode region-specific sulcal variability [13]. They optimized it on the cingulate region, training on 551 subjects from the Human Connectome Project (HCP), with the PCS detection as downstream task. The model's generalizability allowed the detection of a right superior temporal sulcus pattern linked to extreme prematurity [17]. Likewise, using a supervised counterpart of the model, Chavas et al. identified regions correlated to inhibitory control [6].

However, Gaudin et al. did not observe a significant improvement when training on 21,070 subjects from UkBioBank [2] instead of HCP, which calls for a more thorough optimization on UkBioBank to leverage the dataset size. Moreover, their model has been optimized solely for PCS detection. Here, we speculate that this was due to augmentations whose semantic content was not rich enough and to a downstream task that was too simple for optimization. Therefore, we propose the following contributions: by adding as a downstream task a new dataset hand-labeled for complex folding patterns (the orbitofrontal cortex patterns) and by using UkBioBank for training, we develop a new specific augmentation, TrimDepth, and revisit the augmentation pool, the loss, and the backbone. We establish a new state-of-the-art model and prove its ability to detect parallel sulci, their interruptions, and their shape. We name it Champollion V0, after the man who deciphered hieroglyphics, since the present work aims to decipher the hitherto unknown language of cortical folding.

2   Methods

Input Data: Structural MR images of the brain are processed through the BrainVisa Morphologist pipeline[1] that produces a skeletonized negative cast of the brain. It transforms the sulci into surfaces (3D objects of one-voxel width) following the middle of the folding perpendicularly to the brain hull. These

---

[1] https://brainvisa.info.

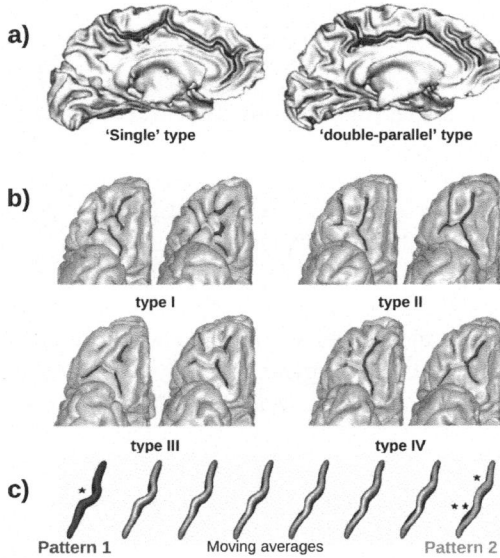

a) 'Single' type   'double-parallel' type

b) type I   type II

type III   type IV

c) Pattern 1   Moving averages   Pattern 2

Fig. 1. The folding patterns studied in this paper. **a)** The two cingulate patterns (blue): the "single" type, with only the cingulate sulcus, and the "double parallel" type, with an additional paracingulate sulcus (PCS) (courtesy from [4]). **b)** Four orbitofrontal cortex pattern types with individual labeled orbitofrontal sulci, from the HCP database. Medial orbital sulcus (MOS) is labeled in red, lateral orbital sulcus (LOS) in blue, intermediate orbital sulcus (IOS) in green, and transverse orbital sulcus (TOS) in yellow. The Type I pattern has a discontinuous MOS and continuous LOS, the Type II a continuous MOS and continuous LOS, the Type III a discontinuous MOS and discontinuous LOS, and Type IV a continuous MOS and discontinuous LOS (courtesy from [24]). **c)**. Moving average of the Isomap first dimension on the central sulcus reveals a continuous change from a single knob (purple) to a double knob (green) pattern (courtesy from [20]). (Color figure online)

cortical skeletons are affinely normalized in the Talairach space with a 2 mm voxel size while constrained to keep a one-voxel width. This preprocessing is meant to emphasize the folding while reducing the bias caused by the original resolution of the acquisition site. Then, a region of interest (ROI) is cropped, using the deep_folding toolbox [7][2], to focus on specific sulci (same tool as Gaudin et al.). Since all datasets are normalized in the same space, a single dataset with hand-labeled sulci is required to define the mask of the ROI. We use a custom dataset (n = 62) independent from the datasets used in this study to define the crops of the cingulate, orbitofrontal, and central regions.

Datasets and Patterns Description: The three datasets used in this paper are the following:

– **UkBioBank** [2] (n = 21070) is a general population cohort used for SSL training.

---

[2] github: neurospin/deep_folding.

- **ACCpatterns** (n = 341) is a dataset composed of subjects taken from [3,5, 9,19,23], with a hand-labeled paracingulate sulcus (PCS, Fig. 1a), our first studied pattern. It is used by Gaudin et al. to optimize the Orig. model, and like Gaudin et al., we focus on the right hemisphere.
- **Human Connectome Project** [26] (HCP). A subset (n = 577) was hand-labeled in the orbitofrontal cortex (OFC) according to the four main interruption patterns (Fig. 1b) [24]. This is our second investigated pattern. We focus on the left hemisphere because the patterns are more uniformly distributed than in the right hemisphere [24]. Their distribution in our data is the following: Type I 49%, Type II 28%, Type III 17%, Type IV 6%. These subjects were also assigned six continuous shape descriptors for their central sulcus using Isomap [22], on a shape similarity matrix [21]. The associated regression tasks are not used during optimization, but only for final evaluation. The first Isomap dimension of the central sulcus is shown in Fig. 1c.

For ACCpatterns and HCP, the data was split into 10 stratified folds according to the label (PCS for ACCpatterns and OFC for HCP), sex, and acquisition sites. Moreover, HCP siblings were systematically assigned to the same splits. The SSL optimization of each task was conducted via 8-fold cross-validation on 8 out of the 10 stratified splits. The remaining 2 splits were retained for the final evaluation. In the SSL setting, ACCpatterns and HCP were only used for the downstream tasks. In the supervised baseline settings (described below), the same 2 splits were utilized for testing, 7 for training and 1 for validation.

Self-supervised Learning Principle and Losses: For each image in a batch, two views are generated using random data augmentations, forming a positive pair. SSL losses bring together the positive pairs in the latent space, ensuring invariance to the augmentations to build semantically expressive representations. To avoid trivial solutions, contrastive methods such as SimCLR consider two views originating from different images as a negative pair and push them away in the latent space [8]. Conversely, BarlowTwins reduces redundancy by decorrelating the latent variables [27]. In this paper, we compare SimCLR, used in the Orig. model, to BarlowTwins.

Although SimCLR is known to perform better with large batch sizes on natural images [8], Gaudin et al. observed a plateau beyond a batch size of 16 [13]. We find the same results when training on a much larger database and hypothesize that the number of latent classes being small, negative pairs in large batches are likely to belong to the same class. Conversely, BarlowTwins is considered a negative-sample-free method [25] and thus looks more suitable.

Hyperparameters: For SimCLR, we use a temperature $\tau = 0.1$ and a batch size of 16 as in the Orig. model. BarlowTwins is reportedly not sensitive to batch size [27]. We set it to 32 as done in [25] with a similar dataset size. According to [27], the regularization hyperparameter λ should be close to $\frac{1}{d}$, d being the latent space size. $\lambda = 10^{-2} (= \frac{2.56}{d}$ with $d = 256)$ is selected after search in the range $(5 * 10^{-3} - 10^{-2})$. The other hyperparameters are taken from the Orig. model: 250 epochs, a learning rate of $4 * 10^{-4}$, and a dropout rate of 5% in the backbone.

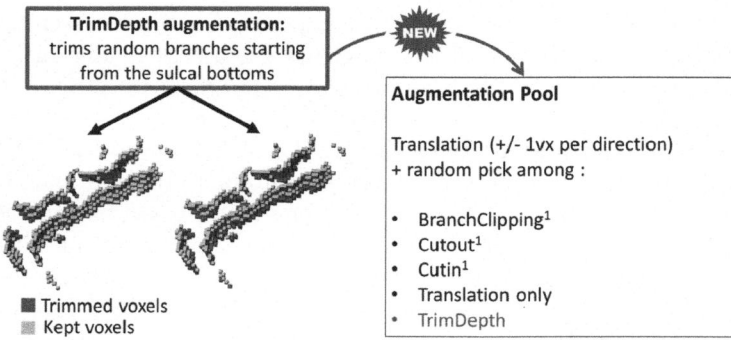

Fig. 2. (Left) Two views examples using TrimDepth. (Right) The augmentation pool. The augmentations marked by [1] have been first defined in [13]

Domain-specific Augmentations: The efficacy of SSL is contingent upon the meticulous design of augmentations [1,8,15] that must maintain the semantics while introducing sufficient variation within the input space. However, there is currently no consensus on defining good augmentations for medical imaging due to the lack of prior knowledge of the semantic content. Consequently, this topic is often overlooked [11].

In the context of cortical skeletons, typical natural image augmentations such as flips or color jittering are not adapted, given that the data is normalized in space and binary. In [13], Gaudin et al. craft domain-specific augmentations for cortical skeletons. Inspired by the popular Cutout [10], they design BranchClipping, which masks skeleton branches until a given percentage of the positive voxels are removed. They also try Cutout and Cutin, which keeps the inside of the mask instead of the outside, but they eventually select BranchClipping as the main augmentation and combine it with a small rotation.

In this paper, we introduce a new augmentation, TrimDepth, which randomly selects folds in the image to trim them of a given depth starting from the bottom. This should help the algorithm focus on sulci shapes rather than depth, which is more relevant. Indeed, sulci shapes are fixed throughout life once the brain is folded but sulci become more shallow with aging. In [3], Cachia et al. define folding patterns as objects invariant with age. Therefore, this augmentation should preserve the subjects' sulcal signatures.

Furthermore, we replace the rotations with small translations (they are faster to compute and don't risk affecting the semantics of images that are not rotation invariant). Differently from Gaudin et al., which look for the augmentation that gives the best result, we hypothesize that using a diverse set of augmentations prevents from learning their individual biases (*e.g.* the PCS may be erased by BranchClipping). We here redesign the augmentation framework so that one augmentation among the following pool is picked for each view with equal probability (except for the translation, which is always applied): BranchClipping, Cutout, Cutin, TrimDepth. By doing a gridsearch, we find that combining all the augmentations yields the best results. Because all the augmentations remove voxels, we do not allow them to be mixed on the same view to prevent the erasure of too many voxels. Additionally, with a probability of 0.2, we apply solely the translation so that the algorithm can learn to represent full images. For Cutout/Cutin, we select a mask covering 30% of the voxels and get the best results when keeping the bottom lines of the masked folds. We hypothesize that maintaining a trace of the global topology helps to rely on the entire image, as opposed to natural images where the semantics are primarily local (*e.g.* an object on a background). For BranchClipping, we find that removing too many branches is detrimental to the OFC classification, so we choose to remove a single branch. But as it may remove a very small number of voxels in some cases, we remove all the bottom voxels aswell. Finally, we select a 2mm depth to be removed by TrimDepth. An illustration of TrimDepth and the new augmentation pool is presented in Fig. 2.

Backbones: We start with the Orig. model's six-layer convolutional network and examine deeper architectures. First, we double the number of layers and filters and increase the initial kernel size from 3 to 7 for 2 mm resolution inputs and to 11 for 1.5 mm resolution. Second, we try a deeper network, the 3D ResNet18 [14] (implemented by [11]). To save computation time, the models using a larger backbone than the original one are trained for 70 epochs, which is found to be enough for convergence.

Model Evaluation: We use the Area Under the Receiver Operating Characteristic Curve (AUC) for classification and the coefficient of determination R^2 for regression. In the multiclass case, we measure the AUCs in a One-vs-Rest scheme and report their weighted average. Deep learning models are trained 5 times to compute a standard deviation. SSL performance is evaluated using a linear model on the latent space (here a linear SVM), the standard way to assess representation quality.

Baselines: The SSL is compared to several baselines of increasing complexity:

- **PCA:** a Principal Component Analysis (PCA) is fit on the UkBioBank skeletons to reduce dimensionality to the latent space size, applied to the datasets of interest, and followed by a linear SVM.
- **Linear models** (Logistic regression, ElasticNet) and **rbf-SVM**: we use these supervised methods on the skeletons as they are classically used to compare with deep learning models in brain MRIs [11]. The regularization and other hyperparameters are detailed in Appendix Table 1.
- **Supervised deep learning:** we tried to either use the SSL augmentation pool or remove the augmentations except for the translation. We obtained best results without the augmentation pool for both classification tasks. We tested the best convolutional network found for SSL and a ResNet18 and found the 12-layer ConvNet backbone to be better for both tasks. We train for 200 epochs and use early stopping to limit overfitting. To handle the OFC class imbalance, we try weighting the loss, but don't achieve better results. The hyperparameter search is detailed in Appendix Table 2.

3   Experiments and Results

SSL Optimization: The Orig. model is here completely redesigned, and the steps are illustrated in Fig. 3. First, we find that increasing the dimension of the latent space (from 10 to 256) is critical for capturing the pattern information in the OFC. Gaudin et al. selected a latent space size of 10 but did not explore values beyond 30, as the parameter did not appear to benefit the PCS detection. Then, the augmentations are replaced with the new pool (Fig. 2). We observe a dramatic increase in performance, especially on the OFC patterns, without additional computational cost. Next, we replace the SimCLR loss with BarlowTwins. It leads to a slight improvement in both tasks and better stability (a standard deviation of 2.5% for SimCLR against 1.0% for BarlowTwins on the OFC patterns), although it does not appear as crucial as the choice of the augmentations.

Furthermore, we investigate larger backbones and achieve a 5% AUC improvement on the OFC patterns (at a non-negligible computational cost).

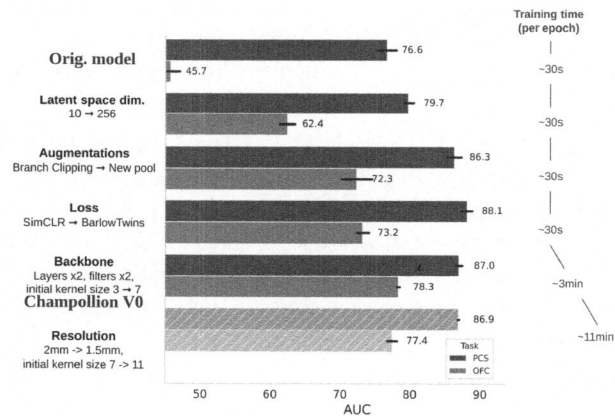

Fig. 3. Summary of the SSL improvements for the PCS and OFC classification. Each model is evaluated with linear SVM using 8-fold cross-validation on the train/validation splits. Training time is reported for a Quadro RTX 5000.

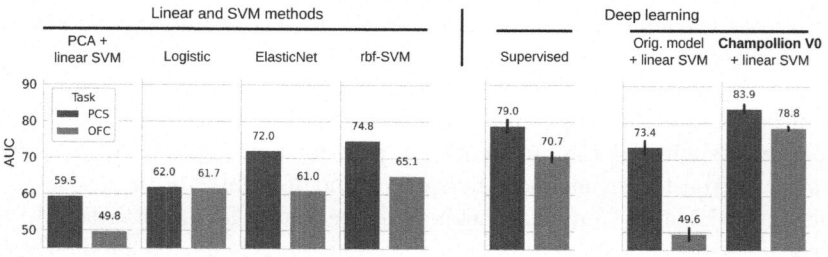

Fig. 4. Comparison between SSL and different baselines on the PCS detection and OFC patterns classification. Each model is evaluated on the test set.

First, we double the number of layers and filters of the original backbone. Subsequently, we observe two distinct performance regimes while training multiple models, with an AUC for OFC detection of approximately 78% for some models and 65% for others, with no obvious difference in the loss value. We hypothesize that the first convolutional layer may be crucial for capturing interruptions before the deeper layers with stride potentially erase them, which may cause instability. Drawing inspiration from ResNet18, we increase the initial kernel size from 3 to 7, which stabilizes the learning, with all models giving an AUC of $\sim$ 78%. We try to further increase the backbone size using ResNet18 but we only achieve random performance. We hypothesize that it may be oversized for our problem.

Finally, we attempt to increase the image resolution to 1.5 mm. We find that the initial kernel must be scaled to 11 for stability, but we do not observe any improvement, while training is significantly slowed down. At first, the results were surprising, as we expected the interruptions to be clearer at higher resolution, leading to a better classification of the patterns. However, a quick visualization indicates that the interruptions at 1.5 mm are still visible at 2 mm.

Champollion V0 outperforms the Orig. model on the test set by 29 points on OFC and 11 points on PCS (AUC, Fig. 4). Although our hyperparameter search may not have been exhaustive, we could not beat SSL with a supervised algorithm. Moreover, all supervised methods failed to classify the type IV OFC pattern (not shown on graph), present in only $\sim$ 6% of the population. By contrast, Champollion V0 performs well on all OFC types (type I: $77.2 \pm 0.6\%$, type II: $83.1 \pm 0.7\%$, type III: $77.4 \pm 0.3\%$, type IV: $81.4 \pm 1.4\%$). Thus, SSL pretraining appears to be particularly valuable for detecting rare patterns.

Application to Isomap Regression: We compare Champollion V0 and the Orig. model in their ability to encode sulcus shape features, which was not assessed by the tasks used for optimization, by performing linear support vector regressions between the latent space of the central sulcus region and the first six dimensions of the central sulcus Isomap. The first dimension is illustrated in Fig. 1c. Champollion V0 encodes all these geometrical features better (Table 1).

Table 1. Comparison of the regression coefficient R^2 between the latent space and the 6 Isomap components, computed on the HCP test set.

Isomap Dim.	1	2	3	4	5	6
Orig. Model	48.2 ± 2.2	42.1 ± 3.2	43.7 ± 12.3	34.5 ± 1.5	0.1 ± 0.1	11.3 ± 2.1
Champollion V0	$\mathbf{54.8 \pm 1.2}$	$\mathbf{58.5 \pm 2.3}$	$\mathbf{68.8 \pm 1.0}$	$\mathbf{56.6 \pm 1.8}$	$\mathbf{26.3 \pm 1.5}$	$\mathbf{38.7 \pm 1.3}$

4 Conclusion and Perspectives

This paper describes Champollion V0, a region-specific encoder that provably detects sulci, their shape, and how they are interrupted. The algorithm achieves

84% AUC on PCS detection and 79% on OFC pattern classification. Given the often ambiguous boundary between classes (*e.g.*, a very short PCS), it is likely that Champollion V0 has reached a plateau for these humanly defined patterns.

We underscore the design of domain-specific augmentations as a cornerstone of our model, while it is largely overlooked in medical imaging [11]. We believe always enriching the augmentation set is the path to further improvements. A possible augmentation would be a lateral trimming of the sulci to highlight the interruptions. Another improvement would be to combine the augmentations differently as large augmentation combinatorics benefit SSL [15].

To further establish Champollion as a foundation model for cortical folding, it will soon be trained on more subjects, 50,000 UkBioBank subjects. Second, we can enforce the latent space to encode biologically relevant patterns using supervision (*e.g.*, pathological patterns) and y-aware loss with metadata [12] (*e.g.*, genetically driven patterns using UkBioBank genetic distances).

We have shown that Champollion V0 bridges the gap between automated and manual labeling of folding patterns. It can now automatically detect known folding patterns for large-scale studies. This has a strong implication regarding the links between folding patterns and clinical endpoints. For example, the OFC patterns have been linked to schizophrenia, but only on small databases [16]. Champollion V0 can now permit the study of such links between folding patterns and schizophrenia on all available schizophrenia datasets.

Acknowledgements. This project has been funded by ANR via FOLDDICO (Project-ANR-20-CHIA-0027), BHT (ProjetIA-22-PESN-0012) and StratifyAging (ProjetIA-22-PESN-0010)

Disclosure of Interests. The authors have no competing interests to declare.

References

1. Balestriero, R., et al.: A cookbook of self-supervised learning. arXiv preprint arXiv:2304.12210 (2023)
2. Bycroft, C., et al.: The UK Biobank resource with deep phenotyping and genomic data. Nature **562**(7726), 203–209 (2018). publisher: Nature Publishing Group
3. Cachia, A., et al.: Longitudinal stability of the folding pattern of the anterior cingulate cortex during development. Dev. Cogn. Neurosci. **19**, 122–127 (2016)
4. Cachia, A., Borst, G., Vidal, J., Fischer, C., Pineau, A., Mangin, J.F., Houdé, O.: The shape of the ACC contributes to cognitive control efficiency in preschoolers. J. Cogn. Neurosci. **26**(1), 96–106 (2014)
5. Chakravarty, M.M., et al.: Striatal shape abnormalities as novel neurodevelopmental endophenotypes in schizophrenia: a longitudinal study. Hum. Brain Mapp. **36**(4), 1458–1469 (2014)
6. Chavas, J., Gaudin, A., Rivière, D., Mangin, J.F.: Regional supervised learning of inhibitory control strength from cortical sulci. In: Medical Imaging with Deep Learning (2024)

7. Chavas, J., Guillon, L., Pascucci, M., Dufumier, B., Rivière, D., Mangin, J.F.: Unsupervised representation learning of cingulate cortical folding patterns. In: Wang, L., Dou, Q., Fletcher, P.T., Speidel, S., Li, S. (eds.) MICCAI 2022. LNCS, vol. 13431, pp. 77–87. Springer, Cham (2022). https://doi.org/10.1007/978-3-031-16431-6_8

8. Chen, T., Kornblith, S., Norouzi, M., Hinton, G.E.: A simple framework for contrastive learning of visual representations. In: International Conference on Machine Learning, pp. 1597–1607 (2020)

9. Delalande, L., et al.: Complex and subtle structural changes in prefrontal cortex induced by inhibitory control training from childhood to adolescence. Dev. Sci. **23**(4), e12898 (2020)

10. DeVries, T., Taylor, G.W.: Improved Regularization of Convolutional Neural Networks with Cutout (2017). arXiv:1708.04552 [cs]

11. Dufumier, B., et al.: Exploring the potential of representation and transfer learning for anatomical neuroimaging: application to psychiatry. Neuroimage **296**, 120665 (2024)

12. Dufumier, B., et al.: Contrastive learning with continuous proxy meta-data for 3D MRI classification. In: de Bruijne, M., et al. (eds.) MICCAI 2021. LNCS, pp. 58–68. Springer, Cham (2021). https://doi.org/10.1007/978-3-030-87196-3_6

13. Gaudin, A., et al.: Optimizing contrastive learning for cortical folding pattern detection. In: Colliot, O., Mitra, J. (eds.) Medical Imaging 2024: Image Processing, vol. 12926, p. 129260Q. International Society for Optics and Photonics, SPIE (2024)

14. He, K., Zhang, X., Ren, S., Sun, J.: Deep residual learning for image recognition. In: 2016 IEEE Conference on Computer Vision and Pattern Recognition (CVPR), pp. 770–778 (2016)

15. Huang, W., Yi, M., Zhao, X., Jiang, Z.: Towards the Generalization of Contrastive Self-Supervised Learning (2023). arXiv:2111.00743 [cs, stat]

16. Isomura, S., et al.: Altered sulcogyral patterns of orbitofrontal cortex in a large cohort of patients with schizophrenia. NPJ Schizophr. **3**, 3 (2017)

17. Laval, J., et al.: Self-supervised contrastive learning unveils cortical folding pattern linked to prematurity. In: Medical Imaging with Deep Learning (2024)

18. Moyal, M., et al.: Orbitofrontal sulcal patterns in catatonia. Eur. Psychiatry **67**(1), e6 (2024)

19. Rapoport, J.L., Gogtay, N.: Childhood onset schizophrenia: support for a progressive neurodevelopmental disorder. Int. J. Dev. Neurosci. Offic. J. Int. Soc. Dev. Neurosci. **29**(3), 251–258 (2011)

20. Sun, Z.Y., Pinel, P., Rivière, D., Moreno, A., Dehaene, S., Mangin, J.F.: Linking morphological and functional variability in hand movement and silent reading. Brain Struct. Funct. **221**(7), 3361–3371 (2016)

21. Sun, Z.Y., et al.: The effect of handedness on the shape of the central sulcus. Neuroimage **60**(1), 332–339 (2012)

22. Tenenbaum, J.B., Silva, V.D., Langford, J.C.: A global geometric framework for nonlinear dimensionality reduction. Science **290**(5500), 2319–2323 (2000)

23. Tissier, C., et al.: Sulcal polymorphisms of the IFC and ACC contribute to inhibitory control variability in children and adults. eNeuro **5**(1), ENEURO.0197–17.2018 (2018)

24. Troiani, V., Snyder, W., Kozick, S., Patti, M.A., Beiler, D.: Variability and concordance of sulcal patterns in the orbitofrontal cortex: a twin study. Psychiatry Res. Neuroimag. **324**, 111492 (2022)

25. Tsai, Y.H.H., Bai, S., Morency, L.P., Salakhutdinov, R.: A note on connecting Barlow twins with negative-sample-free contrastive learning (2021). arXiv:2104.13712 [cs]
26. Van Essen, D.C., Smith, S.M., Barch, D.M., Behrens, T.E.J., Yacoub, E., Ugurbil, K.: The WU-Minn human connectome project: an overview. Neuroimage **80**, 62–79 (2013)
27. Zbontar, J., Jing, L., Misra, I., LeCun, Y., Deny, S.: Barlow twins: self-supervised learning via redundancy reduction. In: International Conference on Machine Learning, pp. 12310–12320 (2021)

Clinical Applications

A Lesion-Aware Edge-Based Graph Neural Network for Predicting Language Ability in Patients with Post-stroke Aphasia

Zijian Chen[1(✉)], Maria Varkanitsa[2], Prakash Ishwar[1], Janusz Konrad[1], Margrit Betke[3], Swathi Kiran[2], and Archana Venkataraman[1]

[1] Department of Electrical and Computer Engineering,
Boston University, Boston, USA
{zijianc,pi,jkonrad,archanav}@bu.edu

[2] Center for Brain Recovery, Boston University, Boston, USA
{mvarkan,skiran}@bu.edu

[3] Department of Computer Science, Boston University, Boston, USA
betke@bu.edu

Abstract. We propose a lesion-aware graph neural network (LEGNet) to predict language ability from resting-state fMRI (rs-fMRI) connectivity in patients with post-stroke aphasia. Our model integrates three components: an edge-based learning module that encodes functional connectivity between brain regions, a lesion encoding module, and a subgraph learning module that leverages functional similarities for prediction. We use synthetic data derived from the Human Connectome Project (HCP) for hyperparameter tuning and model pretraining. We then evaluate the performance using repeated 10-fold cross-validation on an in-house neuroimaging dataset of post-stroke aphasia. Our results demonstrate that LEGNet outperforms baseline deep learning methods in predicting language ability. LEGNet also exhibits superior generalization ability when tested on a second in-house dataset that was acquired under a slightly different neuroimaging protocol. Taken together, the results of this study highlight the potential of LEGNet in effectively learning the relationships between rs-fMRI connectivity and language ability in a patient cohort with brain lesions for improved post-stroke aphasia evaluation.

Keywords: Lesion-aware modeling · Graph neural networks · Functional connectivity · Data augmentation · Aphasia prediction

1 Introduction

Stroke is one of the major causes for disability worldwide [7], with approximately one-third of stroke survivors affected by speech and language impairments, known as aphasia [1]. Resting-state fMRI (rs-fMRI) captures steady-state patterns of co-activation in the brain and provides a unique glimpse into

D. R. Bathula et al. (Eds.): MLCN 2024, LNCS 15266, pp. 91–101, 2025.
https://doi.org/10.1007/978-3-031-78761-4_9

the altered brain network organization due to the stroke [5]. Exploring this relationship is crucial for understanding the mechanisms underlying aphasia and for developing effective, personalized treatment strategies. However, developing models that can simultaneously accommodate patient-specific changes in functional connectivity due to a lesion (i.e., the stroke area) and use this information to predict generalized language impairments remains an open challenge.

Prior studies have attempted to predict language ability using neuroimaging data. The earliest work [17] developed a stacked random forest (RF) model that performed feature selection across multiple modalities and then used these features to predict the composite Aphasia Quotient scored from the revised Western Aphasia Battery, i.e., WAB-AQ [10]. Another study [11] used support vector regression (SVR) to predict WAB-AQ by stacking features from functional MRI, structural MRI, and cerebral blood flow data. Recent work [3] proposed a supervised learning method for feature selection and fusion methods to integrate features from different modalities and predicted WAB-AQ using RF and SVR. An earlier study [2] used similar multimodal ML methods to predict treatment response, rather than baseline functionality. Finally, Wang et al. [21] used persistent diagrams derived from patient rs-fMRI to identify aphasia subtypes. While these studies represent seminal contributions, they largely treat the data as a "bag of features" and do not fully capitalize on network-level information.

We propose to address this gap with Graph neural networks (GNN), which represent the brain as a graph, where nodes correspond to regions of interest (ROIs) and edges represent functional connections between ROIs. Convolutions on the graph aggregate information from neighboring nodes or edges. They can be node-based, as seen in models like BrainGNN [12], GAT [20], and GIN [23], or edge-based, as formulated in the BrainNetCNN [9] and the HGCNN [8] models. GNNs have shown superior performance compared to traditional machine learning techniques in predicting cognitive outcomes related to autism [4,12], aging and intelligence [8], Alzheimer's Disease [25], and ADHD [26]. However, these applications revolve around intact brain networks, which is not the case for a large lesion caused by stroke. Previous work [15,16] took the approach of masking out the lesioned ROIs from the input data. However, this strategy ignores the possibility of informative brain signals from around the lesion boundary. Another challenge is the limited availability of rs-fMRI data from stroke patients. One approach is to reduce the number of features in the analysis [3,11,17]. However, feature selection may inadvertently remove key information in the data, and prior studies have not been diligent about cleanly separating data used for feature selection from that used for performance evaluation [2,17].

In this paper, we introduce a novel lesion-aware edge-based GNN model, which we call LEGNet, that uses rs-fMRI connectivity to predict language ability in patients with post-stroke aphasia. LEGNet is designed to aggregate information from neighboring edges of the brain graph, thus aligning with both the nature of rs-fMRI connectivity and the distributed interactions that contribute to language performance. We incorporate lesion information into LEGNet by encoding the stroke size and position into the model and by using this encoding to constrain the graph convolution process. This frees our approach from masking out the lesioned ROIs, which can lead to information loss. To address data scarcity, we draw from the approach of [14] and develop a comprehensive data augmentation strategy that inserts an "artificial lesion" into healthy neuroimaging data and simulates the corresponding impact on rs-fMRI connectivity and language ability. We demonstrate that LEGNet outperforms baseline deep learning methods on two in-house datasets of patients with post-stroke aphasia.

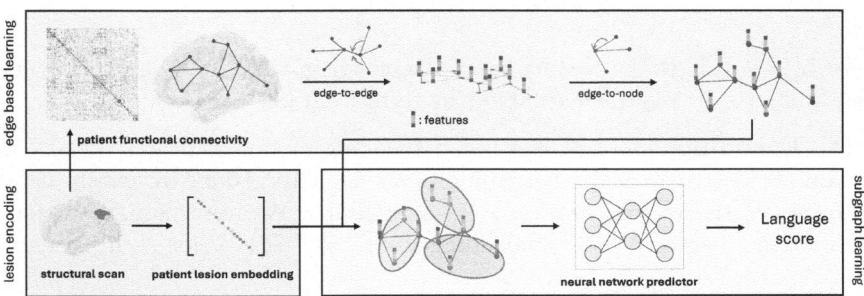

Fig. 1. The Lesion-aware Edge-based BrainGNN Model. **Top:** Edge-to-edge message passing and edge-to-node aggregation. **Bottom Left:** Patient-specific lesion size and position encoding. **Bottom Right:** Subgraph updating and language prediction.

2   Lesion-Aware GNN with Simulated Training Data

2.1   LEGNet Model Architecture

An overview of our LEGNet model architecture is shown in Fig. 1. LEGNet is designed to bridge the gap between the region- or node-based characterization of a lesion and rs-fMRI connectivity, which is defined on edges. As seen, our model includes three components: an edge-based learning module, a lesion encoding module, and a subgraph learning module that connects the two viewpoints.

Formally, let N be the number of ROIs in the brain. The input to LEGNet is the patient rs-fMRI connectivity $\mathbf{X} \in \mathbb{R}^{N \times N}$, which is obtained by exponentiating the correlation matrix computed from the mean time series of non-lesioned voxels within each ROI, as introduced in [15]. If the entire ROI lies within the lesion, then the time series is zero. The entries of $\mathbf{X}$ can be viewed as *edge features* in the underlying brain graph defined on the ROIs.

Edge-Based Learning: From the input $\mathbf{X}$, LEGNet first performs an edge-to-edge convolution [9] given by the following relationship:

$$\mathbf{H}_{ij} = \phi\bigg(\sum_{n \in \mathcal{N}(i)} \mathbf{r}_n \mathbf{X}_{in} + \sum_{n \in \mathcal{N}(j)} \mathbf{c}_n \mathbf{X}_{nj} \bigg), \tag{1}$$

where $\mathbf{H}_{ij} \in \mathbb{R}^{d_0}$ is the feature map of edge (i, j), $\mathcal{N}(i)$ is the set of neighboring nodes to ROI i, including i itself, $\mathbf{r}_n \in \mathbb{R}^{d_0}$ and $\mathbf{c}_n \in \mathbb{R}^{d_0}$ are the learnable filters for each node n, and ϕ is an activation function that is applied element-wise. Intuitively, Eq. (1) aggregates the connectivity information along neighboring edges that share the same end-nodes and updates the edge features accordingly.

Following this step, LEGNet maps the edge features back into the node space:

$$\mathbf{h}_i^{(1)} = \phi\left(\sum_{n \in \mathcal{N}(i)} \mathbf{g}_n \mathbf{H}_{in} + \mathbf{b}_1 \right), \quad i = 1, 2, \dots, N, \tag{2}$$

where $\mathbf{h}_i^{(1)} \in \mathbb{R}^{d_1}$ is the feature map of node i, $\mathbf{g}_n \in \mathbb{R}^{d_1 \times d_0}$ is the learnable filter, and $\mathbf{b}_1 \in \mathbb{R}^{d_1}$ is the learnable bias term from [9].

Lesion Encoding: The LEGNet lesion encoding module captures the size and position of the stroke for downstream processing. This is done by computing the percentage of spared gray matter p_i in each ROI i. We use this information to construct the diagonal lesion embedding matrix $\mathbf{L} \in \mathbb{R}^{N \times N}$ for each patient:

$$\mathbf{L} = \begin{bmatrix} | & | & & | \\ \mathbf{L}_1 & \mathbf{L}_2 & \cdots & \mathbf{L}_N \\ | & | & & | \end{bmatrix} = \begin{bmatrix} p_1 & & \\ & \ddots & \\ & & p_N \end{bmatrix}, \tag{3}$$

If an ROI is intact, then $p_i = 1$ to indicate no lesion; otherwise, $0 \leq p_i < 1$.

Subgraph Learning: At a high level, the subgraph learning module divides the nodes/ROIs into k subgroups based on their lesion encoding information and their (learned) contributions to the final prediction. First, LEGNet uses the lesion encoding $\mathbf{L}$ to update the node representations via:

$$\mathbf{h}_i^{(2)} = \phi\left(\sum_{j \in \mathcal{N}(i)} \mathbf{W}_j \mathbf{h}_j^{(1)} \right), \quad i = 1, 2, \dots, N, \tag{4}$$

where $\mathbf{h}_i^{(2)} \in \mathbb{R}^{d_2}$ is the updated representation for node i, and, inspired by [12], the filters $\mathbf{W}_j \in \mathbb{R}^{d_2 \times d_1}$ are parameterized using the lesion matrix $\mathbf{L}$ as follows:

$$\text{vec}(\mathbf{W}_j) = \mathbf{\Theta}_2 \cdot \psi(\mathbf{\Theta}_1 \mathbf{L}_j) + \mathbf{b}_2. \tag{5}$$

The learnable parameters $\mathbf{\Theta}_2 \in \mathbb{R}^{d_2 d_1 \times k}$ and $\mathbf{\Theta}_1 \in \mathbb{R}^{k \times N}$ are shared across all regions and all subjects. The bias term is $\mathbf{b}_2$, and ψ an activation function.

This mechanism balances the similarity of lesion overlap between regions with their predictive power in the downstream task when learning the weights for each region. The assignment score for each node j is computed as $\psi(\mathbf{\Theta}_1 \mathbf{L}_j)$ and depends on its lesion embedding. The score indicates the involvement of node j in each subgraph. In contrast, the parameter $\mathbf{\Theta}_2$ controls the weight assignment of each subgraph based on the contribution to the prediction. Taken together, ROIs with similar lesion information and functionality are grouped together and updated with similar filters. Following the subgraph learning, the updated node features $\mathbf{h}_i^{(2)}$ are fed into a fully connected layer, with dimension d_3, to predict the scalar WAB-AQ, which quantifies language ability.

Training Loss: We train LEGNet using the mean squared error between the actual y_m and predicted $\hat{y}_m$ language performance for each subject m, together with a ridge regularization term on the network filters:

$$\ell = \frac{1}{M} \sum_{i=1}^{M} (\hat{y}_i - y_i)^2 + \lambda\, R(\mathbf{\Theta}_1, \mathbf{\Theta}_2, \mathbf{b}_1, \mathbf{b}_2, \mathbf{r}, \mathbf{c}, \mathbf{g}), \tag{6}$$

where M is the total number of subjects, R is an L^2-norm, and λ is a hyper-parameter that controls the regularization strength for preventing overfitting.

2.2   Synthetic Data Generation for Model Pre-training

Given the heterogeneity of stroke, we pre-train LEGNet using a large simulated dataset. This pre-trained model is then fine-tuned using our small patient dataset. Our strategy is to insert "artificial lesions" into the neuroimaging data of healthy subjects and simulate its impact on rs-fMRI connectivity and language.

Our pipeline for generating synthetic data is shown in Fig. 2. We first simulate a unique structural lesion for each subject based on the following rules: (1) lesions are left-hemisphere only; (2) lesions are placed randomly but do not cross arterial territories [13]; (3) lesion sizes range from 5% to 20% of one arterial territory; (4) lesions are spatially continuous and simply-connected (i.e., without holes in the inside). Starting from a random seed, the lesion is grown based on a random walk in all directions in the 3D space until it reaches the arterial territory boundary or reaches the threshold size. Holes are filled ad-hoc to meet criterion #4. Next, the artificial lesion is used to mask out voxels when computing ROI mean time series. We also diminish and add Gaussian noise to the connectivity represented in $\mathbf{X}$ between the lesioned region and the rest of the brain, followed by clipping the values to lie within the original connectivity range. Finally, the language performance score is re-scaled proportional to the percentage spared gray matter (< 1) to simulate the negative impact of the lesion on functionality.

2.3   Implementation Details

Simulated-Lesion HCP (HCP-SL): We use rs-fMRI data from 700 randomly selected subjects in the Human Connectome Project (HCP) S1200 database [19] as the foundation for generating synthetic data. Following the standard HCP minimal preprocessing pipeline [18], we parcellate the brain into 246 ROIs using the Brainnetome atlas [6]. The subject language score is accuracy in answering simple math and story-related questions during an fMRI language task. Artificial lesions are inserted and modify the data as described in Sect. 2.2.

Pre-training: Pre-training is done via 10-fold cross validation (CV). We use a two-stage grid search to fix the model hyperparameters $\{\lambda, k, d_0, d_1, d_2, d_3\}$, with

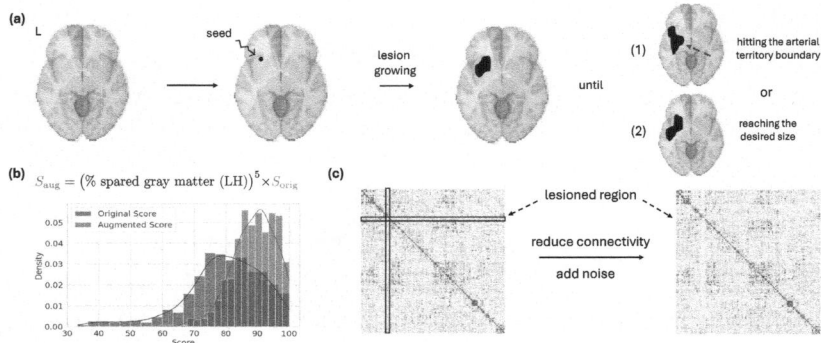

Fig. 2. Synthetic data generation workflow. (a) Artificial lesions are created to lie within a single arterial territory. (b) Language score density is adjusted after the lesion augmentation. (c) Functional connectivity is corrupted in the lesioned area.

a coarse stage used to select a suitable power of 2 from 2^1 to 2^6, followed by a fine stage with increments of 1–2. The regularizer λ is swept across $[10^{-4}, 1]$. The final values are $\{\lambda = 0.005, k = 8, d_0 = 4, d_1 = 8, d_2 = 2, d_3 = 8\}$. We use the Adam optimizer with learning rate starting from 0.01 and decaying by a factor of 0.95 every 20 steps. Early stopping is also applied based on the validation loss. Once the hyperparameters are selected, we pre-train the LEGNet architecture in order to provide a better model initialization for later stages. Given that synthetic lesions may not fully represent the real-world conditions, our pre-training stage is merely designed to provide a better initialization for the stroke dataset.[1]

Application to Post-stroke Aphasia: We use repeated 10-fold CV on our larger in-house dataset (see Sect. 3.1) to evaluate the performance of LEGNet in a real-world setting. The hyperparameters and optimizations are fixed at the values determined during the pre-training phase on synthetic data. The same procedure is applied to all baseline methods. We compare two scenarios: (1) training the models (LEGNet and baselines) from scratch, and (2) using the pre-trained model as the initialization for our repeated CV experiment.

Cross-Dataset Generalization: As further validation, we quantify the language ability prediction performances when the models (LEGNet and baselines) are trained on DS-1 and applied to our second in-house dataset (DS-2) which has slightly different patient characteristics than DS-1.

2.4   Baseline Models

We compare LEGNet with four baselines approaches. The first baseline is a modified BrainGNN model (BrainGNN†) [12], which uses the same subgraph learning modules but does not perform edge-based learning or incorporate lesion information. This serves as an ablation of the edge-based learning and lesion encoding

---

[1] All code and synthetic data are available at https://github.com/zijianch/LEGNet.

modules. The second baseline is BrainNetCNN model [9] with ROIs masked from the rs-fMRI connectivity input if the percentage of spared gray matter is less than 0.3 (BNC-masked). The third baseline is the BrainNetCNN with a two-channel input, with one channel being the unaltered rs-fMRI connectivity and the second channel being a lesion mask (BNC-2channel). These two models serve as an ablation of subgraph learning module. The final baseline is support vector regression (SVR) with the lower-triangle of the rs-fMRI connectivity matrix used as the input feature vector. The baseline models inherit the appropriate subset of hyperparameters from LEGnet. We used the default radial basis function kernel with $C = 100, \gamma = 2/(N(N-1))$, and $\epsilon = 0.1$.

3 Experimental Results

3.1 Datasets of Post-stroke Aphasia

In-House Dataset 1 (DS-1): This dataset consists of 52 patients with chronic post-stroke aphasia with left-hemisphere lesions and aged between 35–80 years. The average time post stroke is approximately 55 months. Structural MRI (T1-weighted; TE = 2.98 ms, TR = 2300 ms, TI = 900 ms, res = 1 mm isotropic) and rs-fMRI (EPI; TE = 20 ms, TR=2/2.4 s, res = $1.72 \times 1.72 \times 3$ mm^3) were acquired on a Siemens 3T scanner. Both scans are pre-processed using the CONN tool-box [22]. Lesion boundaries were delineated manually by trained professionals and normalized to the MNI space. We use the Brainnetome atlas [6] to delineate 246 ROIs for the input rs-fMRI connectivity. Finally, all patients were evaluated using the WAB test [10] to obtain a measure of overall language ability (i.e., WAB-AQ). This value ranges from 0–100 with lower scores indicating severe aphasia, and higher scores indicating mild aphasia.

In-House Dataset 2 (DS-2): This dataset consists of 18 patients with chronic post-stroke aphasia that were recruited separately from DS-1. The average time post stroke is approximately 113 months. While the inclusion criteria and neu-roimaging acquisition protocols are the same as for DS-1, the distribution of WAB-AQ scores is different. This provides an ideal scenario to evaluate cross-dataset generalization of LEGNet and the baseline models.

3.2 Performance Characterization and Model Interpretation

Table 1 reports the predictive performance of each method using repeated 10-fold CV on DS-1. LEGNet achieves the best performance in RMSE, R^2, and correlation coefficient. While it is second-best to BNC-masked in MAE, the difference is not statistically significant. As these methods also serve as ablation models, we notice that all three modules contribute significantly to the final prediction accuracy. As a baseline, we applied the LEGNet architecture to DS-1 from a random initialization, i.e., without having access to synthetic data. To avoid data leakage, we selected the hyperparameters based on the corresponding modules used in previous studies [12,15]. We note a statistically significant

Table 1. Language prediction on DS-1 using repeated 10-fold CV for LEGNet and the baselines in Sect. 2.4. The asterisk * indicates statistically worse performance ($p < 0.05$) compared to the best performing model highlighted in bold. Down-arrow ↓ indicates that lower values are better, and up-arrow ↑ indicates the opposite.

Methods	RMSE ↓	MAE ↓	R^2 ↑	Correlation Coeff. ↑
LEGNet	**17.38 ± 0.44**	12.56 ± 3.02	**0.35 ± 0.03**	**0.61 ± 0.03**
BrainGNN†	19.46 ± 0.30*	14.02 ± 2.59*	0.29 ± 0.04*	0.59 ± 0.02*
BNC-mask	19.18 ± 1.82*	**11.61 ± 2.33**	0.22 ± 0.05*	0.55 ± 0.03*
BNC-2channel	22.78 ± 0.53*	19.41 ± 0.89*	0.21 ± 0.10*	0.53 ± 0.05*
SVR	20.45 ± 0.35*	16.85 ± 0.26*	0.13 ± 0.05*	0.55 ± 0.03*
LEGNet (w/o HCP-SL)	18.39 ± 0.68*	15.33 ± 0.57*	0.29 ± 0.05*	0.58 ± 0.04

decrease in performance w/o HCP, which underscores the importance of using synthetic data to design and initialize the deep network.

Figure 3 (left) illustrates the top two subgraphs identified by LEGNet for the best-performing model during repeated 10-fold CV. The top subgraphs are identified by averaging the subgraph assignment scores for each ROI ($\psi\left(\boldsymbol{\Theta}_1 \mathbf{L}_j\right)$ from Sect. 2.1) across all 52 patients in DS-1. We use Neurosynth [24] to decode the functionality associated with the ROIs assigned to each of the top two subgraphs, as shown in Fig. 3 (right). We note that LEGNet assigns high scores to regions that are related to the language ability. Intuitively, these regions also influence the prediction of language ability, as described in Sect. 2.1.

Table 2. Language prediction on DS-2 with the best-performing model from repeated 10-fold CV on DS-1. Top performance is highlighted in bold.

Methods	RMSE ↓	MAE ↓	R^2 ↑	Correlation Coeff. ↑
LEGNet	**17.71**	**8.74**	**0.19**	**0.44**
BrainGNN†	18.52	12.64	0.11	0.34
BNC-mask	19.24	12.68	0.04	0.31
BNC-2channel	19.70	13.47	0.01	0.28
SVR	18.43	12.65	0.12	0.35
LEGNet (w/o HCP-SL)	18.36	11.26	0.12	0.40

Finally, we tested generalization performance by applying the model that performs best on DS-1 to DS-2 without any fine-tuning (Table 2). In terms of R^2, LEGNet maintains a leading position but, along with the other baselines, also shows a decrease compared to the validation performance in Table 1. While BrainGNN and SVR also show a decrease, BrainNetCNN-based models exhibit a sharper drop, indicating their reduced robustness on unseen data. The other three metrics follow a similar trend. This is expected due to the slight distribution

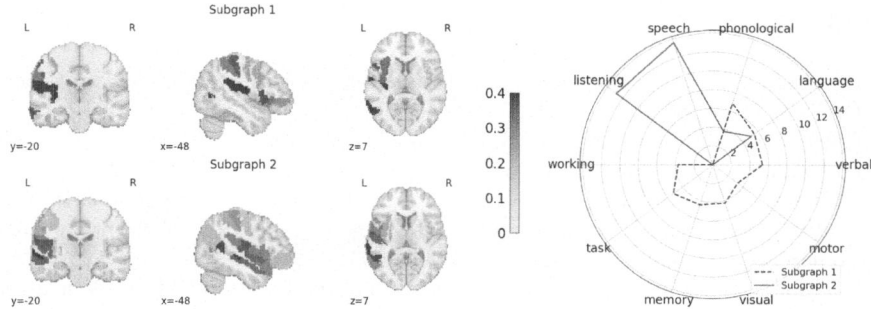

Fig. 3. Left: Top two subgraphs in in DS-1. **Right:** The z−score of each subgraph from Neurosynth [24] indicating the association strength with a particular function.

shift between DS-1 and DS-2. Nevertheless, LEGNet still outperforms all baseline methods, indicating superior generalization ability.

4   Conclusion

We have introduced LEGNet, a novel lesion-aware edge-based graph neural network model designed to predict language performance in post-stroke aphasia patients from rs-fMRI connectivity. LEGNet bridges the gap between the lesion boundary defined on nodes and rs-fMRI connectivity defined on edges, while simultaneously using the lesion size and position to guide both the graph convolution and subgraph identification processes. Our synthetic data generation procedure addresses the challenge of limited patient data by simulating lesioned brain networks in healthy subjects. Pretraining on this synthetic data allows us to perform an unbiased hyperparameter selection and provides a reliable model initialization for fine-tuning on patient data. We demonstrate that LEGNet outperforms state-of-the-art methods in predictive accuracy and generalization ability, thus highlighting its potential as a reliable tool for post-stroke aphasia evaluation. Moreover, since LEGNet does not make assumptions about the specific nature of the lesion, it can be easily adapted to tumors, brain resections, etc. without the need to modify the architecture. Future work will explore attention mechanisms as an alternate means of incorporating the lesion information and exploring the prediction of future (e.g., post-treatment) cognitive states.

Acknowledgements. This work was supported by the National Institutes of Health R01 HD108790 (PI Venkataraman), the National Institutes of Health R01 EB029977 (PI Caffo), the National Institutes of Health R21 CA263804 (PI Venkataraman), the National Institutes of Health P50DC012283 (BU Site PI Kiran) and the National Institutes of Health R01 DC016950 (PI Kiran).

References

1. Berthier, M.L.: Poststroke aphasia: epidemiology, pathophysiology and treatment. Drugs & aging **22**, 163–182 (2005)
2. Billot, A., et al.: Multimodal neural and behavioral data predict response to rehabilitation in chronic poststroke aphasia. Stroke **53**(5), 1606–1614 (2022)
3. Chennuri, S., et al.: Fusion approaches to predict post-stroke aphasia severity from multimodal neuroimaging data. In: Proceedings of the IEEE/CVF International Conference on Computer Vision, pp. 2644–2653 (2023)
4. Dsouza, N.S., Nebel, M.B., Crocetti, D., Robinson, J., Mostofsky, S., Venkataraman, A.: M-GCN: a multimodal graph convolutional network to integrate functional and structural connectomics data to predict multidimensional phenotypic characterizations. In: Medical Imaging with Deep Learning, pp. 119–130. PMLR (2021)
5. Falconer, I., Varkanitsa, M., Kiran, S.: Resting-state brain network connectivity is an independent predictor of responsiveness to language therapy in chronic poststroke aphasia. Cortex (2024)
6. Fan, L., et al.: The human brainnetome atlas: a new brain atlas based on connectional architecture. Cereb. Cortex **26**(8), 3508–3526 (2016)
7. Feigin, V.L., et al.: World stroke organization (WSO): global stroke fact sheet 2022. Int. J. Stroke **17**(1), 18–29 (2022)
8. Huang, J., Chung, M.K., Qiu, A.: Heterogeneous graph convolutional neural network via hodge-laplacian for brain functional data. In: Frangi, A., de Bruijne, M., Wassermann, D., Navab, N. (eds.) IPMI 2023. LNCS, vol. 13939, pp. 278–290. Springer, Cham (2023). https://doi.org/10.1007/978-3-031-34048-2_22
9. Kawahara, J., et al.: BrainNetCNN: convolutional neural networks for brain networks; towards predicting neurodevelopment. Neuroimage **146**, 1038–1049 (2017)
10. Kertesz, A.: Western aphasia battery–revised (2007). https://doi.org/10.1037/t15168-000
11. Kristinsson, S., et al.: Machine learning-based multimodal prediction of language outcomes in chronic aphasia. Hum. Brain Mapp. **42**(6), 1682–1698 (2021)
12. Li, X., et al.: BrainGNN: interpretable brain graph neural network for fMRI analysis. Med. Image Anal. **74**, 102233 (2021)
13. Liu, C.F., et al.: Digital 3D brain MRI arterial territories atlas. Sci. Data **10**(1), 74 (2023)
14. Nandakumar, N., Hsu, D., Ahmed, R., Venkataraman, A.: A deep learning framework to localize the epileptogenic zone from dynamic functional connectivity using a combined graph convolutional and transformer network. In: 2023 IEEE 20th International Symposium on Biomedical Imaging (ISBI), pp. 1–5. IEEE (2023)
15. Nandakumar, N., Manzoor, K., Agarwal, S., Pillai, J.J., Gujar, S.K., Sair, H.I., Venkataraman, A.: Automated eloquent cortex localization in brain tumor patients using multi-task graph neural networks. Med. Image Anal. **74**, 102203 (2021)
16. Nandakumar, N., et al.: A multi-scale spatial and temporal attention network on dynamic connectivity to localize the eloquent cortex in brain tumor patients. In: Feragen, A., Sommer, S., Schnabel, J., Nielsen, M. (eds.) IPMI 2021. LNCS, vol. 12729, pp. 241–252. Springer, Cham (2021). https://doi.org/10.1007/978-3-030-78191-0_19
17. Pustina, D., et al.: Enhanced estimations of post-stroke aphasia severity using stacked multimodal predictions. Hum. Brain Mapp. **38**(11), 5603–5615 (2017)

18. Smith, S.M., et al.: Resting-state fMRI in the human connectome project. Neuroimage **80**, 144–168 (2013)
19. Van Essen, D.C., et al.: The WU-MINN human connectome project: an overview. Neuroimage **80**, 62–79 (2013)
20. Velickovic, P., Cucurull, G., Casanova, A., Romero, A., Lio, P., Bengio, Y., et al.: Graph attention networks. Stat **1050**(20), 10–48550 (2017)
21. Wang, Y., Yin, J., Desai, R.H.: Topological inference on brain networks across subtypes of post-stroke aphasia. ArXiv (2023)
22. Whitfield-Gabrieli, S., Nieto-Castanon, A.: Conn: a functional connectivity toolbox for correlated and anticorrelated brain networks. Brain Connect. **2**(3), 125–141 (2012)
23. Xu, K., Hu, W., Leskovec, J., Jegelka, S.: How powerful are graph neural networks? arXiv preprint arXiv:1810.00826 (2018)
24. Yarkoni, T., Poldrack, R.A., Nichols, T.E., Van Essen, D.C., Wager, T.D.: Large-scale automated synthesis of human functional neuroimaging data. Nat. Methods **8**(8), 665–670 (2011)
25. Zhang, J., Guo, Y., Zhou, L., Wang, L., Wu, W., Shen, D.: Constructing hierarchical attentive functional brain networks for early AD diagnosis. Med. Image Anal. **94**, 103137 (2024)
26. Zhao, K., Duka, B., Xie, H., Oathes, D.J., Calhoun, V., Zhang, Y.: A dynamic graph convolutional neural network framework reveals new insights into connectome dysfunctions in ADHD. Neuroimage **246**, 118774 (2022)

DISARM: Disentangled Scanner-Free Image Generation via Unsupervised Image2Image Translation

Luca Caldera[1], Lara Cavinato[1]([✉]), Andrea Cappozzo[1], Isabella Cama[2],
Sara Garbarino[3], Alessio Cirone[3], Raffaele Lodi[4], Fabrizio Tagliavini[5],
Anna Nigri[6], Silvia De Francesco[7], Francesca Ieva[1],
and RIN-Neuroimaging Network[5]

[1] MOX, Department of Mathematics, Politecnico di Milano, Milan, Italy
lara.cavinato@polimi.it
[2] Department of Mathematics, Università di Genova, Genova, Italy
[3] IRCCS Ospedale Policlinico San Martino, Genova, Italy
[4] Department of Biomedical and Neuromotor Sciences, University of Bologna,
Bologna, Italy
[5] Unit of Neurology (V) and Neuropathology, IRCCS Istituto Neurologico Carlo
Besta, Milan, Italy
[6] Neuroradiology Unit, IRCCS Istituto Neurologico Carlo Besta, Milan, Italy
[7] Laboratory of Neuroinformatics, IRCCS Istituto Centro San Giovanni di Dio
Fatebenefratelli, Brescia, Italy

Abstract. Ensuring reproducibility of Magnetic Resonance (MR) images from different scanners is crucial in multicenter studies, as scanner-induced variability is known to impact the results significantly. To address this problem, we introduce a novel unsupervised deep learning approach aimed at achieving 3 primary objectives/advantages: (1) create a *scanner-free* space that enables the uniform transfer of images across different scanners in a denoised setting, (2) impart the appearance of a specific training scanner to images, transferring its unique characteristics, (3) avoid time-consuming preprocessing of MR images. The proposed methodology is based on disentangling image information into two distinct spaces: one encoding the scanner-specific information and one capturing the anatomical/biological structure of the image. We trained our model on two open-source datasets (ADNI and PPMI) of T1-weighted brain MR images of normal control patients. We tested it

Data used in preparation of this article were obtained from the Alzheimer's Disease Neuroimaging Initiative (ADNI) database (adni.loni.usc.edu). As such, the investigators within the ADNI contributed to the design and implementation of ADNI and/or provided data but did not participate in analysis or writing of this report.

Supplementary Information The online version contains supplementary material available at https://doi.org/10.1007/978-3-031-78761-4_10.

on a real-world clinical dataset from the Italian Neuroimaging Network, comparing its performance with a state-of-the-art model. The results show the superiority of the proposed model in harmonizing images for clinical research, demonstrating its effectiveness in achieving consistent and reproducible harmonization of the MR images across (unseen) scanning environments. Code is available at https://github.com/luca2245/DISARM_Harmonization.

Keywords: Image harmonization · I2I Translation · Magnetic Resonance Imaging · Noise Disentanglement · Scanner-free Imaging

1   Introduction

With the rapid growth of brain Magnetic Resonance Imaging (MRI) datasets collected from various centers, the power of data-derived insights is becoming increasingly fundamental in understanding brain-related diseases, which can significantly enhance medical practice by providing robust statistical evidence. However, these centers often follow unharmonized acquisition protocols, leading to substantial variability in the extracted biomarkers. MRI scanners can indeed produce images with different contrast, brightness, and spatial characteristics due to variations in hardware, software, calibration, and other factors, including discrepancies that arise from machine maintenance, protocols, environmental conditions, and operator expertise. In multicenter studies involving diverse scanners and centers, this variability can confound results, highlighting the critical need for harmonized images to ensure robust analysis [19,22,27].

Several methodologies have been proposed for harmonizing images and specifically multicenter MRI data. Image-to-image (I2I) translation family includes methods that aim to map an input one into an output image while preserving certain semantic properties. Specifically, these generative models produce translated images, that appear to be drawn from the distribution of the target domain. They are categorized according to the type of supervision, i.e., supervised (directional or bidirectional) or unsupervised (cyclic consistency, autoencoders, disentangler), or to the type of translation, i.e., one-to-one, one-to-many, many-to-many [1]. Transformers [6] are models based on self-attention mechanism with the power of capturing long-range context dependencies. Still, they have been shown to overlook important high-frequency features that may be the site for biology- and pathology-related information [24]. Style transfer is a fully unsupervised deep learning framework [11], wherein image harmonization is treated as a style transfer problem rather than a domain transfer problem.

Only few attempts have been brought out in the medical domain, among all NeuroHarmonize [15], Calamiti [26], MURD [12] and ComBat [16,23]. At the crossroad between style transfer and I2I translation families, StarGANv2 [4] stands out as a single framework that addresses both the diversity of generated images and scalability across multiple domains, so far demonstrating significantly improved results compared to baseline methods. While showing promising results in the harmonization task, this approach necessitates extensive preprocessing, including skull-stripping of the T1 images [18], nonuniformity correction [20], and registration to a standard space, making the end-to-end process heavy and intricate.

In this work, with the aim of overcoming inter-scanner variability of images to allow findings to be translated into clinical practice, we present DISARM, a novel model to harmonize 3D MR images by disentangling anatomical structure and scanner-specific information. The model is designed to produce *scanner-free* images while preserving the original anatomical structure and biologically informative data as to achieve robust generalizability across various scanners.

2   Proposed Methodology

We propose a harmonization model for 3D medical images extending upon DRIT++ network architecture [10], with specific application to T1-weighted brain MRI data. Particularly, our model targets the scanner-free generation of 3D MR images to clean out clinical images from batch effects.

2.1   Mathematical Formulation

Let $\mathcal{X} = \bigcup_{i=1}^{N} \mathcal{X}_i \in \mathbb{R}^{1 \times H \times W \times D}$ be a set of MR images and denote with $\mathcal{X}_i$ the set of images acquired from the $i - th$ scanner domain among N distinct domains. We assume that the images can be disentangled into two latent spaces $(\mathcal{B}, \mathcal{S})$, being $\mathcal{B}$ the space encoding the information related to the brain anatomical structure and being $\mathcal{S}$ the scanner space, specifically targeting the scanner effects. An image x drawn from $\mathcal{X}_i$ can be thus obtained as the combination of $\mathcal{B}$ and $\mathcal{S}$. The scanner-free harmonization task consists of noising out the scanner-specific information by plugging a random Gaussian noise $\mathcal{N}(0,1)$ instead of the $\mathcal{S}$ distribution. The final objective is to ensure robust generalizability across various scanners, including those not encountered during the training phase, and to deliver a model that does not necessitate time-consuming preprocessing of MR images.

2.2   Architecture

The backbone of the architecture includes two encoders and a generator. The brain structure encoder $E_b : \mathcal{X} \to \mathcal{B}$ maps an input image $x \in \mathcal{X}$ into a lower-dimensional space $\mathcal{B}$ encoding the information related to the brain structure while the scanner encoder $E_s : (\mathcal{X}, \mathcal{C}) \to \mathcal{S}$ takes an input image $x \in \mathcal{X}$ and its associated scanner label within the space $\mathcal{C}$ and operates as a variational autoencoder, producing a parametric distribution of the scanner effect. The scanner effect latent vector z_i^s is thus sampled as:

$$z_i^s = \sigma_i \cdot \varepsilon + \mu_i; \quad z_i^s \in \mathcal{S}_i; \quad \varepsilon \sim \mathcal{N}(0,1).$$

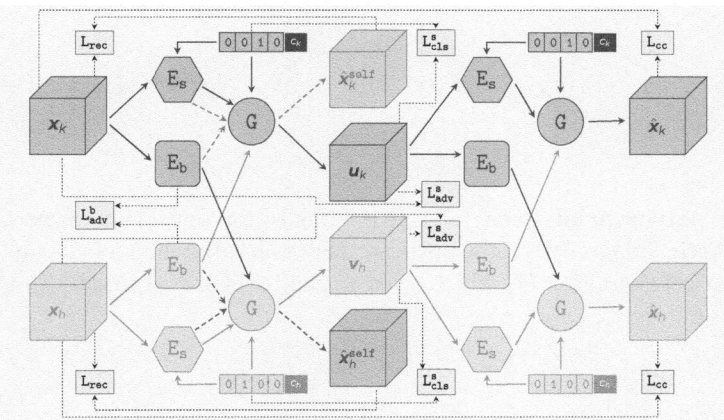

Fig. 1. High level functional diagram of DISARM network's training procedure. Discriminators, L_{lat}, and L_{KL} have not been displayed due to space reasons.

The generator $G : (\mathcal{B}, \mathcal{S}, \mathcal{C}) \rightarrow \mathcal{X}$ produces an image $x \in \mathcal{X}$ that preserves a specified brain structure within the space $\mathcal{B}$ while incorporating a scanner attribute from the space $\mathcal{S}$ associated to its label within the space $\mathcal{C}$. To transfer a scanner attribute of scanner $\mathcal{X}_h$ to another image $x_k \in \mathcal{X}_k$, the scanner latent space attribute z_h^s is plugged into the generator together with its label $c_h \in \mathcal{C}_h$ and the brain attribute $z_{x_k}^b$, obtaining the image x_k as it was acquired by the scanner h. Upon this strategy, we can achieve a scanner-free harmonization of images by sampling the scanner effect latent vector from $\varepsilon \sim \mathcal{N}(0, 1)$ and plugging it into G together with input x. The scanner-free space is denoted as $\mathcal{F}$ and the encoded scanner label is defined as a vector of zeros. Figures explaining the procedures are provided in the Supplementary Materials.

2.3   Losses

The training process is shown in Fig. 1 and is carried out on pairs of images (x_k, c_k) and (x_h, c_h) from randomly selected scanner domains (K and H) among a pool of N domains. Each image is simultaneously fed into its own E_b and E_s, yielding the corresponding latent representations $z_l^s = E_s(x_l, c_l)$; $z_{x_l}^b = E_b(x_l)$ for $l = \{h, k\}$. The training follows the minimization of a set of losses governing the different subnetworks in the architecture with reference to Fig. 1.

Brain Structure Adversarial Loss (L_{adv}^b)**:** It aims to ensure that the brain encoders E_b generate scanner-independent brain embeddings, that is, map images into a shared space $\mathcal{B}$ where no domain membership is distinguishable. To do this, the latent representations $z_{x_k}^b$ and $z_{x_h}^b$ are fed into a discriminator D_b which tries to identify their domain membership as follows:

$$L_{\text{adv}}^{\text{b}} = \frac{1}{2}\,\mathbb{E}_{\boldsymbol{x}_k}\left[\log\Big[D_b(\boldsymbol{z}_{\boldsymbol{x}_k}^b)(1-D_b(\boldsymbol{z}_{\boldsymbol{x}_k}^b))\Big]\right] + \frac{1}{2}\,\mathbb{E}_{\boldsymbol{x}_h}\left[\log\Big[D_b(\boldsymbol{z}_{\boldsymbol{x}_h}^b)(1-D_b(\boldsymbol{z}_{\boldsymbol{x}_h}^b))\Big]\right]$$

Scanner Adversarial Loss $(L_{\text{adv}}^{\text{s}})$: It aims to force the generators G to produce realistic images. To this purpose, the generated images $\boldsymbol{u}_k$ and $\boldsymbol{v}_h$ are fed into a discriminator D_s which tried to distinguish real images to generated images as follows:

$$L_{\text{adv}}^{\text{s}} = \frac{1}{2}\,\mathbb{E}_{\boldsymbol{x}_k}\left[\log\Big[D_s(\boldsymbol{x}_k)(1-D_s(\boldsymbol{u}_k))\Big]\right] + \frac{1}{2}\,\mathbb{E}_{\boldsymbol{x}_h}\left[\log\Big[D_s(\boldsymbol{x}_h)(1-D_s(\boldsymbol{v}_h))\Big]\right]$$

Scanner Classification Loss $(L_{\text{cls}}^{\text{s}})$: The discriminator D_c is trained to predict the scanner label of the generated images, to ensure that its scanner-related information is compliant with the desired one (i.e., the one plugged in):

$$L_{\text{cls}}^{\text{s}} = \mathbb{E}_{\boldsymbol{x}_k}\left[-\log\big[D_c(\boldsymbol{c}_k|\boldsymbol{u}_k)\big]\right] + \mathbb{E}_{\boldsymbol{x}_h}\left[-\log\big[D_c(\boldsymbol{c}_h|\boldsymbol{v}_h)\big]\right]$$

Cycle Consistency Loss (L_{cc}): To maintain cross-cycle consistency, $\boldsymbol{u}_k$ and $\boldsymbol{v}_h$ undergo a backward translation process where they are re-generated in their own original scanner domain. The reconstructed images $\hat{\boldsymbol{x}}_k = G(E_b(\boldsymbol{v}_h), E_s(\boldsymbol{u}_k), \boldsymbol{c}_k)$ and $\hat{\boldsymbol{x}}_h = G(E_b(\boldsymbol{u}_k), E_s(\boldsymbol{v}_h), \boldsymbol{c}_h)$ need to be as maximally similar to the original $\boldsymbol{x}_k$ and $\boldsymbol{x}_h$ in terms of:

$$L_{cc} = \mathbb{E}_{\boldsymbol{x}_k,\boldsymbol{x}_h}\left[\big\|\hat{\boldsymbol{x}}_k - \boldsymbol{x}_k\big\| + \big\|\hat{\boldsymbol{x}}_h - \boldsymbol{x}_h\big\|\right]$$

Self Reconstruction Loss (L_{rec}): It ensures that the reconstructed image $\hat{\boldsymbol{x}}_l^{\text{self}} = G(E_b(\boldsymbol{x}_l), E_s(\boldsymbol{x}_l), \boldsymbol{c}_l)$ closely matches the original image $\boldsymbol{x}_l$ for $l = h, k$. To do this, $\boldsymbol{x}_k$ and $\boldsymbol{x}_h$ are passed through the encoders (E_s, E_b) and the generator G and thus reconstructed as $\hat{\boldsymbol{x}}_k^{\text{self}}$ and $\hat{\boldsymbol{x}}_h^{\text{self}}$. The loss thus computes:

$$L_{\text{rec}} = \mathbb{E}_{\boldsymbol{x}_k,\boldsymbol{x}_h}\left[\big\|\hat{\boldsymbol{x}}_k^{\text{self}} - \boldsymbol{x}_k\big\| + \big\|\hat{\boldsymbol{x}}_h^{\text{self}} - \boldsymbol{x}_h\big\|\right]$$

Total Loss. The overall model loss is defined as follows:

$$L_{\text{tot}} = \lambda_{\text{cc}}L_{\text{cc}} + \lambda_{\text{rec}}L_{\text{rec}} + \lambda_{\text{lat}}L_{\text{lat}} + \lambda_{\text{KL}}L_{\text{KL}} - \lambda_{\text{adv}}^{\text{b}}L_{\text{adv}}^{\text{b}} - \lambda_{\text{cls}}^{\text{s}}L_{\text{cls}}^{\text{s}} - \lambda_{\text{adv}}^{\text{s}}L_{\text{adv}}^{\text{s}}$$

where L_{lat} serves to force the mean vector $\boldsymbol{\mu}_i$ extracted by the scanner encoder E_s to be close to a standard Gaussian, and L_{KL} is the Kullback-Leibler loss which aims to align the scanner effects representation with a prior Gaussian distribution. Values for parameters are provided in Supplementary Materials.

3   Experimental Setting

Datasets. We validated our approach using T1-weighted MR images from normal controls aged 60 to 80 years, with a slice thickness of 1 mm, obtained on a 3T scanner. In particular, we trained DISARM on a dataset comprising 335 images from Alzheimer's Disease Neuroimaging (ADNI3 study) [7] and Parkinson's Progression Marker Initiatives (PPMI) [13] datasets. Training images came from $N = 5$ different scanners (163 from `Prisma Fit` (**PrF**), 64 from `Prisma`, 30 from `Achieva dStream`, 38 from `Skyra`, and 40 from `TrioTim`). We tested the results on a clinical dataset from the Italian Neuroimaging Network (RIN), which we will refer to as the RIN dataset. This dataset comprises 117 images acquired with 7 scanners (17 from `Achieva dStream` (**Ac**), 20 from `DISCOVERY MR750` (**DI**), 23 from `Ingenia CX` (**In**), 40 from `Prisma` (**Pr**), 2 from `Signa HDxt` (**Si**), 12 from `Skyra` (**Sk**), 3 from `Skyra Fit` (**SkF**)). The cohort's age range is 49 to 80, with an average age of 64.5423 ± 8.6295.

Preprocessing. All the images were (1) normalized to the standard orientation to facilitate further steps, (2) applied a bias-field correction to mitigate the magnetic field variations in the MRI scanners [9,21], and (3) registered to the reference Standard MNI152-T1-1mm [8]. Moreover, we exploited a data augmentation procedure through dense random elastic deformation [14]. Implementation details are given in Supplementary Material.

Benchmarking. To benchmark our model, we evaluated the performances of the pre-trained StarGANv2 [4] on the RIN dataset. The required preprocessing included that the images were skull-stripped [18] and corrected for nonuniformity using the N3 method [20] in Freesurfer. Furthermore, the images needed to be linearly registered to a standard MNI template (ICBM 152 Nonlinear Symmetric atlas [5]) and resized to isotropic 1 mm^3 voxels. According to StarGANv2 implementation, harmonization was performed on a sliding window of 3 image's slices with a stride of 1. The final T1 harmonized MRI 3D volumes were then reconstructed by combining these harmonized partial volumes. As StarGANv2 model do not allow for scanner-free harmonization, the `Prisma Fit` scanner space was used as reference to be compared with the proposed model.

Evaluation Procedure. Three experiments were employed to test and compare models' performance. First, we evaluated whether the proposed model could accurately attribute the effects of one scanner to images captured by another. We thus compared grey level distributions of the output images harmonized on Prisma Fit scanner space by DISARM and StarGANv2. Second, we tested the differences in grey level distributions of the output images harmonized on $\mathcal{F}$ by DISARM. In both cases, we employed K-sample Anderson-Darling test [17]. Lastly, we evaluated the predictive performance in classifying patients' ages, to test the power of the harmonization model to enhance the biological information. We employed a regression model fed with the total gray volumes TGV [2,3] extracted from the RIN MR images before and after harmonization. We included the interactions with scanner variable, evaluating the significance of the scanner information in terms of p-values of the regression β coefficients. For this analysis, "before harmonization" also refers to before the preprocessing pipeline.

4   Results

The evaluation process involves harmonizing the 117 normal controls from the RIN dataset into both the Prisma Fit scanner space, using DISARM and StarGANv2, and the scanner-free space $\mathcal{F}$, using only DISARM. In Supplementary Materials, we provide example output images for both models.

4.1   Evaluation of Prisma Fit scanner information transfer

To test the harmonization task, we calculated the average grayscale distribution across all images for each scanner, before and after transferring the images to the Prisma Fit scanner space. We accounted resulting images from both the proposed model and StarGANv2. In Fig. 2, we display the boxplot of the grey-level distributions per scanner-specific image sets before and after harmonization by DISARM and StarGANv2. Of note, the harmonization of DISARM relies on the whole target scanner domain, whereas StarGANv2 relies on a single image as a reference. Therefore, the reference distribution (shown in gray) for DISARM is the average gray distribution of images acquired with Prisma Fit, whereas the StarGANv2 reference distribution is the one of a selected Prisma Fit image. The two plots (a) and (b) display results for DISARM, showing a massive alignment of the distributions after harmonization (p = 0.39). The two plots (c) and (d) show results for StarGANv2. Of note, (a) and (c) highlight the outcomes of the different preprocessing methods used for DISARM and StarGANv2, where the latter is much stronger. In (d), we notice how output distribution do not perfectly align with the reference image (p = 0.03).

4.2   Evaluation of Scanner Free Space $\mathcal{F}$

We tested the harmonization performance in the scanner-free space $\mathcal{F}$. In Fig. 3, we compare the average gray distributions for each scanner before and after the transfer to $\mathcal{F}$. We observe how the average post-harmonization distributions tend to align across all scanners, indicating successful harmonization (p = 0.36).

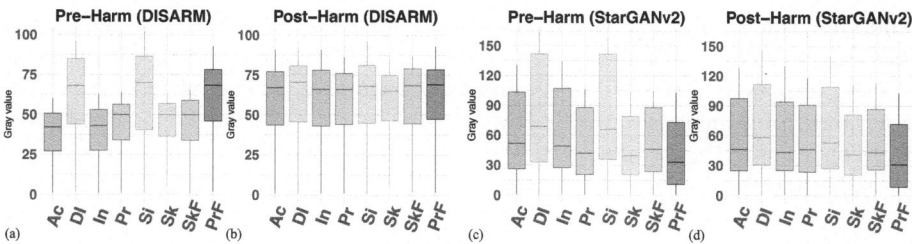

Fig. 2. Average gray distributions with Average Pool in each scanner pre- and post-transfer to `Prisma Fit` scanner space. DISARM result: (a) distributions after Preprocessing and (b) distributions post-harmonization; StarGAN result: (c) distributions after Preprocessing and (d) distributions post-harmonization. (Color figure online)

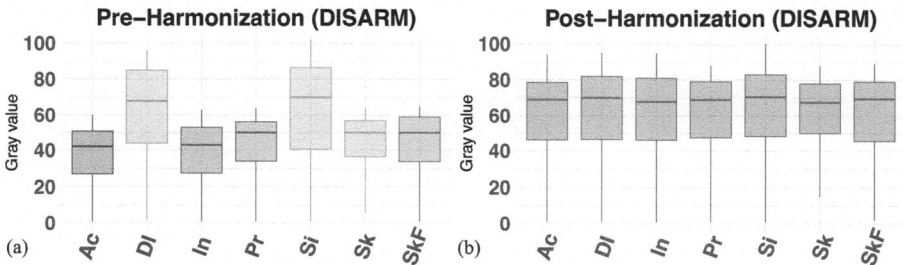

Fig. 3. The average gray distributions in each scanner before (a) and after (b) the transfer to Scanner-Free space $\mathcal{F}$. (Color figure online)

4.3 Age Prediction Task

In Table 1, we present the results of the regression models before and after harmonization for both DISARM and StarGANv2 models. Resulting regression lines are displayed in Supplementary Materials. Before harmonization, interactions with different scanners are significant ($p < 0.05$). After DISARM harmonization, the interaction with various scanners is no longer significant ($p > 0.05$), whereas after StarGANv2 harmonization, it remains weakly significant for two scanners ($p < 0.10$). This reflects the findings of the first experiment, highlighting how the scanner-related information still impact on the age regression task even after the harmonization procedure and sensibly contribute to the prediction. The Root Mean Squared Error (RMSE) before harmonization is 7.19. After harmonization with DISARM, it is 7.05, and after harmonization with StarGANv2, it is 7.45.

4.4 Preservation of Anatomical Structure

To evaluate the preservation of anatomical structures after harmonization, we use the Structural Similarity Index Measure (SSIM). We focus on the structural comparison between the original image and its harmonized version since we lack a ground truth for comparing also contrast and luminance [25]. For a set of 100 random image pairs from the RIN dataset, which exhibit different anatomical structures, the average SSIM value is 0.70 ± 0.07. The average SSIM between all 117 test images of the RIN dataset and their harmonized versions in the scanner-free space, obtained using DISARM, is 0.96 ± 0.007. When harmonization is performed using the Prisma Fit scanner as reference, the average SSIM value is 0.955 ± 0.008.

Table 1. Significance of Total Gray Volume (TGV) and TGV-scanner interactions in age regression models before and after DISARM and StarGANv2.

Variables	Pre-Harmonization P-value	DISARM P-value	StarGANv2 P-value
(Intercept)	**<0.0001**	**<0.0001**	**<0.0001**
TGV	**<0.0001**	**<0.0001**	**<0.0001**
TGV-Achieva	**0.0321**	0.381	**0.078**
TGV-DISCOVERY	**0.0827**	0.101	0.332
TGV-Ingenia	**0.0170**	0.522	**0.081**
TGV-Skyra	0.8977	0.686	0.901

5   Conclusions

We introduced a novel method to harmonize T1-weighted MR images from various scanners, operating across the entire 3D volume to account for spatial correlations among different brain structures during reconstruction. Our approach allows for image transfer from different scanners in two distinct ways: (1) transferring images to a scanner-free space, ensuring consistent appearances regardless of the original scanner source; (2) mapping images to the space of one of the scanners used in the model's training, embedding the unique characteristics of the selected scanner into the transferred image. The model demonstrated robust generalization also to scanners not included in the training data, though the relatively small size of our training dataset indicates room for improvement with more data. Additionally, our method eliminates the need for time-consuming preprocessing steps like skull-stripping, which can be flawed and might remove parts of the brain or leave non-brain tissue remnants. This is particularly beneficial for applications requiring analysis of the entire head image, such as studies involving head trauma or skull deformities. Future expansions of the proposed approach would be carried out using travelling heads and include: (1) eliminating the bias field correction step to further streamline the preprocessing pipeline; (2) improve the model by integrating custom losses aimed at enhancing the fit for the scanner-free space; (3) improving the preservation of brain structure in harmonized images; and (4) extending the model to other imaging modalities and anatomical sites.

Acknowledgments. This study was funded by the Italian Minister of Health (RCR; RRC-2016-2361095; RRC-2017-2364915; RRC-2018-2365796; RCR-2019-23669119_001; RCR 2020-23670067; RCR 2022-23682285) and the Ministry of Economy and Finance (CCR-2017-23669078).

L. Cavinato is funded by the National Plan for NRRP Complementary Investments - project n. PNC0000003 - AdvaNced Technologies for Human-centrEd Medicine (project acronym: ANTHEM).

The authors acknowledge the support by MUR grant Dipartimento di Eccellenza 2023-2027.

Disclosure of Interests. The authors have no competing interests to declare that are relevant to the content of this article.

References

1. Alotaibi, A.: Deep generative adversarial networks for image-to-image translation: a review. Symmetry **12**(10), 1705 (2020)
2. Bauer, T., et al.: Subcortical grey matter volume and asymmetry in the long-term course of Rasmussen's encephalitis. Brain Commun. **5**(6), fcad324 (2023)
3. Charroud, C., Turella, L.: Subcortical grey matter changes associated with motor symptoms evaluated by the unified Parkinson's disease rating scale (Part III): a longitudinal study in Parkinson's disease. NeuroImage: Clinical **31**, 102745 (2021)
4. Choi, Y., Uh, Y., Yoo, J., Ha, J.W.: Stargan v2: diverse image synthesis for multiple domains. In: Proceedings of the IEEE/CVF Conference on Computer Vision and Pattern Recognition, pp. 8188–8197 (2020)
5. Fonov, V.S., Evans, A.C., McKinstry, R.C., Almli, C.R., Collins, D.: Unbiased nonlinear average age-appropriate brain templates from birth to adulthood. Neuroimage **47**, S102 (2009)
6. Guo, Z., Gu, Z., Zheng, B., Dong, J., Zheng, H.: Transformer for image harmonization and beyond. IEEE Trans. Pattern Anal. Mach. Intell. (2022)
7. Jack Jr, C.R., et al.: The Alzheimer's disease neuroimaging initiative (ADNI): MRI methods. J. Magnet. Resonance Imaging Off. J. Int. Soc. Magnet. Reson. Med. **27**(4), 685–691 (2008)
8. Jenkinson, M., Bannister, P., Brady, M., Smith, S.: Improved optimization for the robust and accurate linear registration and motion correction of brain images. Neuroimage **17**(2), 825–841 (2002)
9. Jenkinson, M., Beckmann, C.F., Behrens, T.E., Woolrich, M.W., Smith, S.M.: Fsl. Neuroimage **62**(2), 782–790 (2012)
10. Lee, H.Y.: Drit++: diverse image-to-image translation via disentangled representations. Int. J. Comput. Vision **128**, 2402–2417 (2020)
11. Liu, M., et al.: Style transfer generative adversarial networks to harmonize multisite MRI to a single reference image to avoid overcorrection. Hum. Brain Mapp. **44**(14), 4875–4892 (2023)
12. Liu, S., Yap, P.T.: Learning multi-site harmonization of magnetic resonance images without traveling human phantoms. Commun. Eng. **3**(1), 6 (2024)
13. Marek, K., et al.: The Parkinson progression marker initiative (PPMI). Prog. Neurobiol. **95**(4), 629–635 (2011)
14. Pérez-García, F., Sparks, R., Ourselin, S.: Torchio: a python library for efficient loading, preprocessing, augmentation and patch-based sampling of medical images in deep learning. Comput. Methods Programs Biomed. **208**, 106236 (2021)

15. Pomponio, R., et al.: Harmonization of large MRI datasets for the analysis of brain imaging patterns throughout the lifespan. Neuroimage **208**, 116450 (2020)
16. Radua, J., et al.: Increased power by harmonizing structural MRI site differences with the combat batch adjustment method in enigma. Neuroimage **218**, 116956 (2020)
17. Scholz, F.W., Stephens, M.A.: K-sample anderson-darling tests. J. Am. Stat. Assoc. **82**(399), 918–924 (1987)
18. Ségonne, F., et al.: A hybrid approach to the skull stripping problem in MRI. Neuroimage **22**(3), 1060–1075 (2004)
19. Shinohara, R.T., et al.: Volumetric analysis from a harmonized multisite brain MRI study of a single subject with multiple sclerosis. Am. J. Neuroradiol. **38**(8), 1501–1509 (2017)
20. Sled, J.G., Zijdenbos, A.P., Evans, A.C.: A nonparametric method for automatic correction of intensity nonuniformity in MRI data. IEEE Trans. Med. Imaging **17**(1), 87–97 (1998)
21. Smith, S.M., et al.: Advances in functional and structural MR image analysis and implementation as FSL. Neuroimage **23**, S208–S219 (2004)
22. Takao, H., Hayashi, N., Ohtomo, K.: Effect of scanner in longitudinal studies of brain volume changes. J. Magn. Reson. Imaging **34**(2), 438–444 (2011)
23. Torbati, M.E., et al.: A multi-scanner neuroimaging data harmonization using ravel and combat. Neuroimage **245**, 118703 (2021)
24. Wang, P., Zheng, W., Chen, T., Wang, Z.: Anti-oversmoothing in deep vision transformers via the fourier domain analysis: from theory to practice. arXiv preprint arXiv:2203.05962 (2022)
25. Wang, Z., Bovik, A.C., Sheikh, H.R., Simoncelli, E.P.: Image quality assessment: from error visibility to structural similarity. IEEE Trans. Image Process. **13**(4), 600–612 (2004)
26. Zuo, L., et al.: Unsupervised MR harmonization by learning disentangled representations using information bottleneck theory. Neuroimage **243**, 118569 (2021)
27. Zuo, X.N., Xu, T., Milham, M.P.: Harnessing reliability for neuroscience research. Nat. Hum. Behav. **3**(8), 768–771 (2019)

Segmenting Small Stroke Lesions with Novel Labeling Strategies

Liang Shang, Zhengyang Lou, Andrew L. Alexander, Vivek Prabhakaran, William A. Sethares, Veena A. Nair, and Nagesh Adluru$^{(\boxtimes)}$

University of Wisconsin-Madison, Madison, WI 53706, USA
adluru@wisc.edu

Abstract. Deep neural networks have demonstrated exceptional efficacy in stroke lesion segmentation. However, the delineation of small lesions, critical for stroke diagnosis, remains a challenge. In this study, we propose two straightforward yet powerful approaches that can be seamlessly integrated into a variety of networks: **M**ulti-**S**ize **L**abeling (MSL) and **D**istance-**B**ased **L**abeling (DBL), with the aim of enhancing the segmentation accuracy of small lesions. MSL divides lesion masks into various categories based on lesion volume while DBL emphasizes the lesion boundaries. Experimental evaluations on the Anatomical Tracings of Lesions After Stroke (ATLAS) v2.0 dataset showcase that an ensemble of MSL and DBL achieves consistently better or equal performance on recall (**3.6%** and **3.7%**), F1 (**2.4%** and **1.5%**), and Dice scores (**1.3%** and 0.0%) compared to the top-1 winner of the 2022 MICCAI ATLAS Challenge on both the subset only containing small lesions and the entire dataset, respectively. Notably, on the mini-lesion subset, a single MSL model surpasses the previous best ensemble strategy, with enhancements of **1.0**% and **0.3**% on F1 and Dice scores, respectively. Our code is available at: https://github.com/nadluru/StrokeLesSeg.

Keywords: Small Lesion · Segmentation · Data Augmentation · U-Net · Stroke Lesion · MRI

1 Introduction

Strokes, characterized by an inadequate blood supply to specific brain regions, rank as the second leading cause of death and third leading cause of disability globally [3]. Accurate segmentation of stroke lesions helps clinicians to better diagnose and evaluate any treatment risks [13]. For instance, segmentation aids in identifying the location and extent of the ischemic core and penumbra, which are crucial for determining the eligibility and efficacy of thrombolysis or thrombectomy [1]. Additionally, segmentation can help monitor the evolution of the lesion over time and assess the response to treatment.

Supplementary Information The online version contains supplementary material available at https://doi.org/10.1007/978-3-031-78761-4_11.

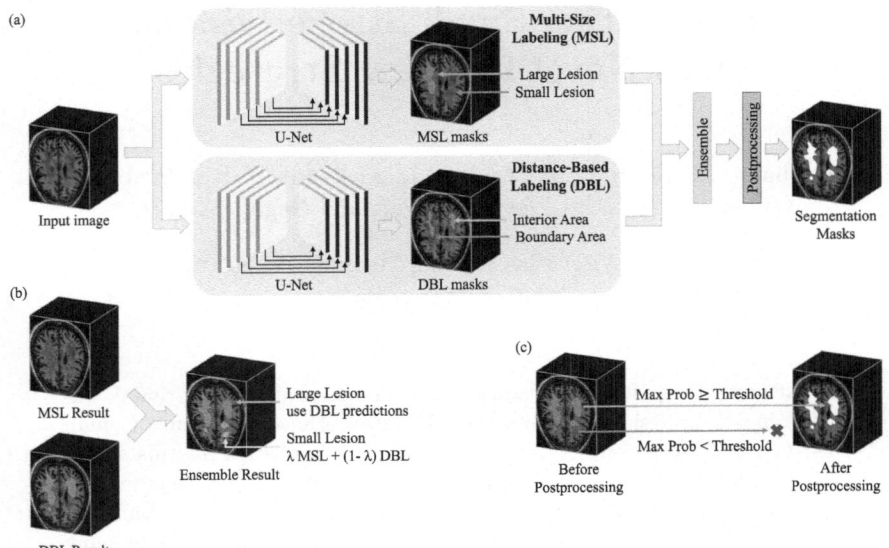

Fig. 1. (a) Overview of our method. We first propose <u>M</u>ulti-<u>S</u>ize <u>L</u>abeling (MSL) and <u>D</u>istance-<u>B</u>ased <u>L</u>abeling (DBL). These two labeling strategies accentuate small lesions and lesion boundaries, respectively. Then, we ensemble models trained with MSL and DBL. Following, post-processing is applied to generate the final segmentation mask. **(b) Proposed ensemble strategy.** A linear interpolation between MSL and DBL results is applied for small lesions to generate ensemble results. For large lesions, we exclusively rely on DBL results. **(c) Proposed postprocessing strategy.** The postprocessing pipeline is designed to enhance segmentation accuracy by filtering out small lesions whose maximum probability falls below a predefined threshold.

The segmentation of small stroke lesions has even more significant clinical implications. Accurate segmentation and quantification of them can provide valuable information for stroke diagnosis and prognosis and for reducing the risk of progressing to post-stroke vascular contributions to cognitive impairment and dementia (VCID) [2, 14]. However, small stroke lesions are often overlooked or misdiagnosed by radiologists due to their low contrast and subtle appearance. It is thus essential to develop methods that can accurately segment small lesions to assist the radiologists.

While deep neural networks have demonstrated impressive performance in detecting [17] and segmenting [6] stroke lesions from Magnetic Resonance Imaging (MRI) scans, they still fail short in segmenting small lesions. We argue that this is because most methods assess their performance based on the match between the number of voxels correctly/incorrectly detected. Since large-volume lesions contain many more voxels than small ones, measures such as the overall Dice score are inherently biased towards accurately detecting large lesions. Accordingly, delineating small lesions remains challenging, despite their critical importance for stroke diagnosis.

Among the existing literature, only a handful of studies focus on segmenting small brain lesions exclusively from brain MRIs. For example, [7,10,16] develop ensemble learning strategies that combine the outputs of multiple CNN blocks operating at different resolutions, which often require specific model architectures. SPiN [15] explores the use of subpixel embedding to retain fine-grained details of the input, particularly in segmenting 2D MRI slices, and its application to 3D MRIs is limited due to the computational complexity associated with image super-resolution techniques.

To this end, we introduce two innovative labeling strategies, **M**ulti-**S**ize **L**abeling (MSL) and **D**istance-**B**ased **L**abeling (DBL), designed to categorize the binary segmentation mask into multiple classes based on the lesion volume and distance to the non-lesion region, respectively. To the best of our knowledge, our study is the first to address the often overlooked and clinically important problem of small lesion segmentation from a data augmentation perspective. Implemented without modifying the feature extractor in the U-Net [12] architecture, MSL and DBL empower the network to differentiate between lesion types with distinct features, thus emphasizing small lesions and the boundary region, respectively. Our contributions can be summarized as follows:

- Our novel labeling strategies (MSL and DBL) are tailored to enhance the performance of small stroke lesion segmentation. Both strategies can be seamlessly integrated into a wide range of segmentation models.
- We propose supplementary ensemble and postprocessing strategies that effectively leverage the advantages of MSL and DBL.
- On the Anatomical Tracings of Lesions After Stroke (ATLAS) v2.0 dataset [8], our ensemble strategy consistently outperforms the leading baseline ensemble in terms of recall, F1, and Dice scores across both the subset specifically targeting small lesions and the entire dataset. Moreover, a single MSL model surpasses the best baseline ensemble on the mini-lesion subset.

2   Approach

Typical state-of-the-art segmentation studies on stroke lesions [4,7,10,15,16] employ binary segmentation masks that distinguish only between normal and lesion regions. We note that due to the inherent imbalance in lesion sizes, it may be advantageous to categorize lesions explicitly based on either their volume or their proximity to the non-lesion region. While this approach maintains the same feature extractor architecture as standard models, it encourages the final classifier to discern different lesion types based on these distinct features. To this end, we introduce **M**ulti-**S**ize **L**abeling (MSL) and **D**istance-**B**ased **L**abeling (DBL), aimed at enhancing the learning process for small lesion. These methods transform the voxel-wise binary classification task for stroke lesion segmentation into a voxel-wise multi-way classification task, facilitating more nuanced and effective segmentation but only adding 0.05% additional FLOPs and 0.007% additional parameters to the network. The segmentation pipeline of our methods is illustrated in Fig. 1.

2.1   Proposed Label Strategies

Multi-size Labeling (MSL). Segmenting small lesions poses a challenge due to the extreme volume imbalance of lesions of different sizes. In the Anatomical Tracings of Lesions After Stroke (ATLAS) v2.0 dataset [8], for instance, among 655 training images, 517 lesions have a volume of less than 100 voxels, totaling approximately 50,000 voxels. On the other hand, a single larger lesion in the dataset can exceed 100,000 voxels. Consequently, disregarding small lesions in predictions does not significantly impact measures of segmentation quality, such as the Dice score. To emphasize the segmentation of small lesions, we introduce $\underline{M}$ulti-$\underline{S}$ize $\underline{L}$abeling (MSL), which categorizes lesion voxels based on their volumes. Specifically, for a voxel k belonging to a lesion K, we assign it to one of the following categories

$$\text{lesion voxel } k \in \begin{cases} \text{tiny lesion,} & \text{if } |K| < 100, \\ \text{small lesion,} & \text{if } 100 \leq |K| < 1,000, \\ \text{medium lesion,} & \text{if } 1,000 \leq |K| < 10,000, \\ \text{large lesion,} & \text{if } |K| \geq 10,000, \end{cases} \tag{1}$$

where $|K|$ represents the volume of the lesion. While balancing the number of voxels for MSL is not feasible due to the extreme variability in lesion sizes, employing this logarithmic categorization ensures each class contains a roughly equal number of lesions, $i.e.$, within the same order of magnitude. We also explore alternative categorizations in our ablation studies.

Distance-Based Labeling (DBL). Interior regions within lesions and normal tissues tend to lack important features for segmentation. In contrast, boundary voxels may contain important information as they represent edges between the lesion and non-lesion regions. Hence, an alternative labeling strategy aims to delineate boundaries from interior regions. We call this strategy $\underline{D}$istance-$\underline{B}$ased $\underline{L}$abeling (DBL). DBL categorizes lesion voxels k as

$$\text{lesion voxel } k \in \begin{cases} \text{boundary region,} & \text{if distance to non-lesion region} \leq 2, \\ \text{interior region,} & \text{if distance to non-lesion region} > 2, \end{cases} \tag{2}$$

to ensure a relatively narrow boundary region with the number of voxels in each class remains within the same order of magnitude. Several alternative categorizations of MSL and DBL are explored in our ablation studies. For completeness of this work, the detailed lesion distribution across categories for both MSL and DBL are listed in Sec. A of the supplementary materials.

Generating Binary Segmentation Masks. As the binary segmentation task transitions into a multi-class voxel-level classification task, additional steps are required to derive a binary segmentation mask from the output. Generally, the network logits are fed into a softmax function to compute the predicted probabilities $p_{k,i}$ of each voxel k belonging to each class $i \in \{0, 1, 2...\}$. In our case, the class 0 represents the background class, and the remaining classes denote the foreground classes assigned by MSL or DBL. The likelihood of a voxel k being part of a lesion, p_k, can be obtained by aggregating the probabilities of all foreground classes by $p_k = \sum_{i \geq 1} p_{k,i}$. Afterwards, the segmentation mask can be achieved by thresholding all voxels with $p_k > 0.5$.

2.2 Ensemble and Postprocessing Pipelines

Ensemble. While MSL significantly enhances the network's capability to segment small lesions, its performance in segmenting larger lesions may be compromised due to the inherent complexity of the multi-class classification task. Similarly, although DBL adeptly highlights the boundaries of both small and large lesions, its efficacy may diminish when segmenting tiny lesions as their boundary and interior regions can be hard to distinguish. To circumvent these limitations, we propose a novel ensemble strategy that integrates both MSL and DBL to ensure a balanced enhancement in the accuracy of lesion segmentation across a diverse range of sizes, thereby addressing a critical gap in current segmentation methodologies. This approach first calculates the probabilities of a voxel k being a lesion separately from the MSL and DBL models, denoted $p_{k,\mathrm{MSL}}$ and $p_{k,\mathrm{DBL}}$, respectively. Then, as illustrated in Fig. 1b, for a lesion prediction K identified as tiny or small, *i.e.*, $|K| < 1000$, we linearly interpolate between $p_{k,\mathrm{MSL}}$ and $p_{k,\mathrm{DBL}}$. Conversely, for larger lesions, we rely solely on $p_{k,\mathrm{DBL}}$. Consequently, the final likelihood of a voxel k being part of a lesion is

$$p_k = \begin{cases} \lambda \cdot p_{k,\mathrm{MSL}} + (1 - \lambda) \cdot p_{k,\mathrm{DBL}}, & \text{if } |K| < 1000, \\ p_{k,\mathrm{DBL}}, & \text{otherwise,} \end{cases} \tag{3}$$

where λ is the mixing rate determined through ablation studies.

Postprocessing (PP). The postprocessing (PP) pipeline aims to further reduce segmentation errors for small lesions. As shown in Fig. 1c, for each prediction of a tiny or small lesion K with $|K| < 1000$, if the maximum probability among all lesion voxels, $\max_{k \in K} p_k$, is less than a pre-defined probability threshold p_t determined through ablation studies, we exclude it from the segmentation mask.

3 Experimental Setting

Dataset. We demonstrate the effectiveness of our proposed labeling strategies using the Anatomical Tracings of Lesions After Stroke (ATLAS) v2.0 dataset [8]. This dataset comprises several training and testing subsets, with a training subset of 655 T_1-weighted MRIs are public with corresponding lesion segmentation masks. All the images in the dataset were corrected for intensity non-uniformity, intensity standardized, and linearly registered to the MNI space, where each voxel corresponds to a volume of 1 mm^3. Within the 655 MRI scans in the training dataset, the lesion volumes span from 13 to over 200,000 voxels, with 138 MRI scans exclusively containing lesions with volumes less than 1000, constituting a subset of mini-lesions.

Table 1. Main results. The bottom 3 rows are our results while the top 5 rows are baselines from [4]. Results in **bold** represent the better one between the baseline ensemble and our ensemble strategies. Compared to the baseline ensemble, which secured first place in the 2022 MICCAI ATLAS challenge, our ensemble result (MSL+DBL) consistently achieves better or equal performance on recall (**3.6%** and **3.7%**), F1 (**2.4%** and **1.5%**), and Dice scores (**1.3%** and 0.0%) on both the mini-lesion subset and the entire dataset, respectively.

Method	Mini-Lesion Subset				Entire Dataset			
	Dice	F1	Precision	Recall	Dice	F1	Precision	Recall
Default	0.417	0.500	0.509	0.628	0.635	0.549	0.686	0.561
DTK10	0.416	0.491	0.550	0.597	0.629	0.547	0.697	0.560
Res U-Net	0.433	0.502	0.481	0.667	0.638	0.540	0.644	0.578
Self-Training	0.434	0.510	0.497	0.661	0.647	0.550	0.672	0.578
Ensemble	0.443	0.574	0.627	0.618	**0.645**	0.575	**0.747**	0.553
MSL	0.446	0.584	0.734	0.612	0.632	0.566	0.727	0.559
DBL	0.421	0.536	0.655	0.576	0.634	0.581	0.766	0.556
MSL+DBL	**0.456**	**0.598**	**0.699**	**0.654**	**0.645**	**0.590**	0.721	**0.590**

Implementation. Our models are developed using PyTorch [11] and nnU-Net [5]. The models are trained for 1000 epochs, with Dice together with cross-entropy (CE) as the compound loss, a batch size of 2, SGD as the optimizer, an initial learning rate of 0.01, a momentum of 0.99, and z-score normalization. During training, two image patches of dimensions $128 \times 128 \times 128$ are extracted from each of the two MRI images in the batch to serve as inputs to the model. One patch is randomly cropped from the original MRI, while the other is specifically cropped to ensure that lesion voxels are centered within the patch. This approach balances the diversity of training samples with a focus on the critical regions containing lesions. Across all training schemes, we conducted a size-balanced 5-fold cross-validation using the data split outlined in [4], whereby the training dataset is evenly partitioned into 5 folds, ensuring nearly identical lesion volume distributions within each fold. Our results are summarized across the 5-fold models over both the entire dataset (655 scans) and the mini-lesion subset (138 scans). For evaluation. we employed four metrics, Dice, F1, precision, and recall scores, to evaluate performance. The Dice score is computed **voxel-wise**, while the latter three scores are computed **lesion-wise**.

4   Results

The evaluation of MSL, DBL, and our optimal ensemble results are illustrated in Table 1. To establish a solid baseline, we replicated the training of the top-performing model from the 2022 MICCAI ATLAS Challenge [4]. This baseline employs a standard U-Net [12] trained with binary segmentation masks and optimized using the Dice+CrossEntropy compound loss. Building on this foundation, we implemented and assessed our proposed MSL and DBL methods.

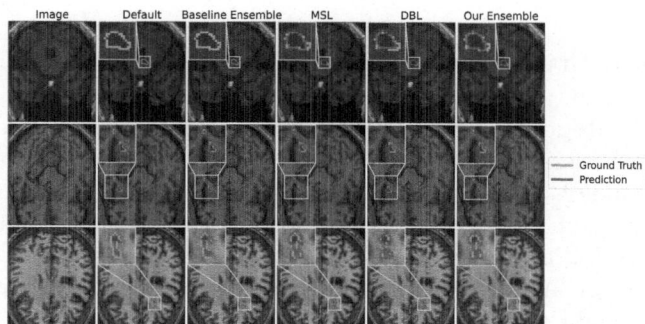

Fig. 2. Visualization of segmentation results. The green contours represent the ground truth, while the red contours depict the predicted lesions. These segmentation results indicate that our methods accurately label more lesions that are missed, *i.e.*, false negative, by the baselines.

For evaluating small lesion detection, while we can report the small lesion detection performance of our method on the entire dataset, it is not feasible to accurately compare to baseline methods which are not explicitly size-aware. For example, these methods might segment a single large lesion, subsuming nearby small lesions that are disconnected in reality/ground truth, thus complicating the exclusion of large lesions. To address this, we selected a pure subset of 138 images exclusively containing lesions with volumes less than 1,000 voxels, denoted as the mini-lesion subset, to accurately compare the core performance of detecting small lesions. On this subset, MSL surpasses the best ensemble results in [4], which include models trained under four different schemes: the default setting, a substitution of the TopK10 loss function (DTK10), a switch to the more intricate Res U-Net architecture, and a self-training scheme leveraging an additional 300 testing scans. MSL achieves higher precision (**10.7%**), F1 (**1.0%**), and Dice (**0.3%**) scores, showcasing its effectiveness in detecting small lesions. Conversely, DBL outperforms their Default and DTK10 counterparts on both the mini-lesion subset and the entire dataset, which utilize the same network architecture. This outcome underscores the effectiveness of accentuating the boundary region.

Moreover, our optimal ensemble, which includes four training schemes including MSL and DBL models trained on both Dice+CrossEntropy and Dice+Focal [9] losses, consistently achieves better or equal performance on recall (**3.6%** and **3.7%**), F1 (**1.3%** and **1.5%**), and Dice scores (**1.3%** and 0.0%) compared to the baseline ensemble results on both the mini-lesion subset and the entire dataset, respectively. Figure 2 offers qualitative segmentation results obtained from MSL, DBL, our ensemble method, and comparison with the schemes proposed in [4]. These results suggest our methods have a superior ability to segment small lesions. For completeness, we have detailed the performance of MSL and DBL models trained on Dice+Focal losses, along with all post-processing hyperparameters in Sec. B and C of the supplementary materials, respectively.

5   Ablation Studies

Number of Categories in MSL and DBL. For MSL, we explore the separation of binary lesion masks into 3, 4, or 5 lesion classes with varying sizes, as shown in Table 2. Notably, splitting into 4 categories yields the most favorable performance, underscoring the significance of striking a balance between the advantages of isolating small lesions and the complexity associated with multiclass classification. As for DBL, we compare the outcomes of splitting solely into boundary and interior regions versus incorporating an additional transition region. Retaining only a relatively narrow boundary region achieves optimal performance, signifying the pivotal role of the lesion boundary in the segmentation process. We also provide the configuration and lesion distribution of each method in Sec. A of the supplementary materials.

Table 2. Ablation studies on the number of categories for each labeling strategy. Categorizing the binary segmentation masks into lesions with 4 different sizes yields the best MSL performance while separating into only the boundary and interior regions achieved the best DBL performance.

Method	Mini-Lesion Subset				Entire Dataset			
	Dice	F1	Precision	Recall	Dice	F1	Precision	Recall
Default	0.417	0.500	0.509	0.628	0.635	0.549	0.686	0.561
DTK10	0.416	0.491	0.550	0.597	0.629	0.547	0.697	0.560
Res U-Net	0.433	0.502	0.481	0.667	0.638	0.540	0.644	0.578
Self-Training	0.434	0.510	0.497	0.661	0.647	0.550	0.672	0.578
Ensemble	0.443	0.574	0.627	0.618	**0.645**	0.575	**0.747**	0.553
MSL	0.446	0.584	0.734	0.612	0.632	0.566	0.727	0.559
DBL	0.421	0.536	0.655	0.576	0.634	0.581	0.766	0.556
MSL+DBL	**0.456**	**0.598**	**0.699**	**0.654**	**0.645**	**0.590**	0.721	**0.590**

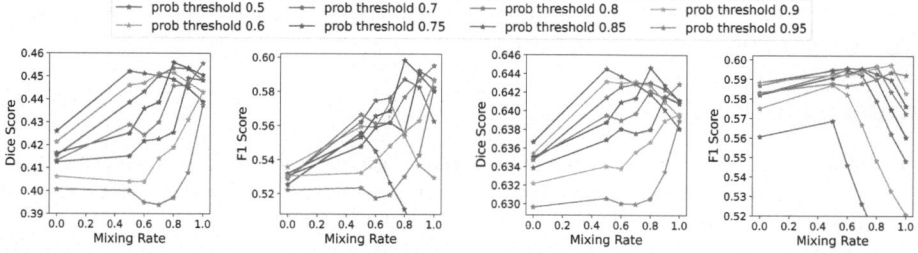

(a) Dice and F1 on the mini-lesion subset.   (b) Dice and F1 on the entire dataset.

Fig. 3. Ablation studies of mixing rate and postprocessing threshold. For Dice and F1 scores, higher is better. A mixing rate of 0.8 with a post-processing threshold of 0.75 is found as the optimal configuration.

Mixing Rate and Postprocessing (PP) Threshold in Ensemble. Our ensemble and postprocessing (PP) strategies require pre-set hyperparameters, including probability thresholds and mixing rates. Figure 3 illustrates the Dice and F1 scores corresponding to various PP thresholds and mixing rates, evaluated on both the mini-lesion subset and the entire dataset. Among these configurations, a mixing rate of 0.8 paired with a PP threshold of 0.75 yields the most favorable average ranking, thus establishing our optimal setting. Notably, the Dice score shows consistent improvement with an increasing mixing rate for PP thresholds ¿0.6, highlighting the advantage of relying more on MSL for segmenting small lesions. The F1 score exhibits consistent behavior as well: it improves until a mixing rate of 0.5, then declines when evaluated across the entire dataset but continues to improve for the mini-lesion subset for PP thresholds ¿0.6.

6 Conclusion

This study introduces Multi-Size Labeling (MSL) and Distance-Based Labeling (DBL) methodologies, compatible with a variety of segmentation networks, aimed at enhancing the segmentation accuracy of small stroke lesions by emphasizing small lesions and the boundaries of the lesions, respectively. In tackling this challenge, our proposed ensemble strategy consistently achieves better or equal performance on recall (3.6% and 3.7%), F1 (2.4% and 1.5%), and Dice scores (1.3% and 0.0%) compared to the ensemble result from the top-1 winner of the 2022 MICCAI ATLAS Challenge on both the mini-lesion subset and the entire dataset, respectively. Furthermore, a single MSL model outperforms the baseline ensemble strategy with improvements of 1.0% and 0.3% on F1 and Dice scores, respectively. These findings underscore the effectiveness of MSL and DBL in stroke lesion segmentation, particularly in small stroke lesion segmentation.

Acknowledgments. We would like to acknowledge the support from NIH grants R01NS123378, P50HD105353, NIH R01NS105646, NIH R01NS11102, and R01NS117568.

Disclosure of Interests. The authors have no competing interests to declare that are relevant to the content of this article.

References

1. Borst, J., et al.: Effect of extended CT perfusion acquisition time on ischemic core and penumbra volume estimation in patients with acute ischemic stroke due to a large vessel occlusion. PLoS One **10**(3), e0119409 (2015)
2. Corriveau, R.A., et al.: The science of vascular contributions to cognitive impairment and dementia (VCID): a framework for advancing research priorities in the cerebrovascular biology of cognitive decline. Cell. Mol. Neurobiol. **36**, 281–288 (2016)
3. Feigin, V.L., et al.: Pragmatic solutions to reduce the global burden of stroke: a world stroke organization-lancet neurology commission. Lancet Neurol. **22**(12), 1160–1206 (2023)
4. Huo, J., et al.: MAPPING: Model average with post-processing for stroke lesion segmentation. arXiv preprint arXiv:2211.15486 (2022)
5. Isensee, F., Jaeger, P.F., Kohl, S.A., Petersen, J., Maier-Hein, K.H.: nnU-Net: a self-configuring method for deep learning-based biomedical image segmentation. Nat. Methods **18**(2), 203–211 (2021)
6. Ito, K.L., Kim, H., Liew, S.L.: A comparison of automated lesion segmentation approaches for chronic stroke T1-weighted MRI data. Hum. Brain Mapp. **40**(16), 4669–4685 (2019)
7. Kamnitsas, K., et al.: Efficient multi-scale 3D CNN with fully connected CRF for accurate brain lesion segmentation. Med. Image Anal. **36**, 61–78 (2017)
8. Liew, S.L., et al.: A large, curated, open-source stroke neuroimaging dataset to improve lesion segmentation algorithms. Sci. Data **9**(1), 320 (2022)
9. Lin, T.Y., Goyal, P., Girshick, R., He, K., Dollár, P.: Focal loss for dense object detection. In: Proceedings of the IEEE International Conference on Computer Vision, pp. 2980–2988 (2017)
10. Liu, C.F., et al.: Deep learning-based detection and segmentation of diffusion abnormalities in acute ischemic stroke. Commun. Med. **1**(1), 61 (2021)
11. Paszke, A., et al.: PyTorch: an imperative style, high-performance deep learning library. In: Advances in Neural Information Processing Systems, vol. 32 (2019)
12. Ronneberger, O., Fischer, P., Brox, T.: U-Net: convolutional networks for biomedical image segmentation. In: Medical Image Computing and Computer-Assisted Intervention–MICCAI 2015: 18th International Conference, Munich, Germany, October 5-9, 2015, Proceedings, Part III 18, pp. 234–241. Springer (2015)
13. Tsai, J.Z., et al.: Automated segmentation and quantification of white matter hyperintensities in acute ischemic stroke patients with cerebral infarction. PLoS ONE **9**(8), e104011 (2014)
14. Wardlaw, J.M., et al.: Neuroimaging standards for research into small vessel disease and its contribution to ageing and neurodegeneration. Lancet Neurol. **12**(8), 822–838 (2013)
15. Wong, A., et al.: Small lesion segmentation in brain MRIs with subpixel embedding. In: International MICCAI Brainlesion Workshop, pp. 75–87. Springer (2021)
16. Xu, B., et al.: Orchestral fully convolutional networks for small lesion segmentation in brain MRI. In: 2018 IEEE 15th International Symposium on Biomedical Imaging (ISBI 2018), pp. 889–892. IEEE (2018)
17. Zhang, S., Xu, S., Tan, L., Wang, H., Meng, J.: Stroke lesion detection and analysis in MRI images based on deep learning. J. Healthc. Eng. **2021**, 1–9 (2021)

A Progressive Single-Modality to Multi-modality Classification Framework for Alzheimer's Disease Sub-type Diagnosis

Yuxiao Liu[1], Mianxin Liu[2], Yuanwang Zhang[1], Kaicong Sun[1],
and Dinggang Shen[1,3,4(✉)]

[1] School of Biomedical Engineering and State Key Laboratory of Advanced Medical
Materials and Devices, ShanghaiTech University, Shanghai 201210, China
[2] Shanghai Artificial Intelligence Laboratory, Shanghai 200232, China
[3] Shanghai United Imaging Intelligence Co., Ltd., Shanghai 200230, China
[4] Shanghai Clinical Research and Trial Center, Shanghai 201210, China
dgshen@shanghaitech.edu.cn

Abstract. The current clinical diagnosis framework of Alzheimer's disease (AD) involves multiple modalities acquired from multiple diagnosis stages, each with distinct usage and cost. Previous AD diagnosis research has predominantly focused on how to directly fuse multiple modalities for an end-to-end one-stage diagnosis, which practically requires a high cost in data acquisition. Moreover, a significant part of these methods diagnose AD without considering clinical guideline and cannot offer accurate sub-type diagnosis. In this paper, by exploring inter-correlation among multiple modalities, we propose a novel progressive AD sub-type diagnosis framework, aiming to give diagnosis results based on easier-to-access modalities in earlier low-cost stages, instead of all modalities from all stages. Specifically, first, we design 1) a text disentanglement network for better processing tabular data collected in the initial stage, and 2) a modality fusion module for fusing multi-modality features separately. Second, we align features from modalities acquired in earlier low-cost stage(s) with later high-cost stage(s) to give accurate diagnosis without actual modality acquisition in later-stage(s) for saving cost. Furthermore, we follow the clinical guideline to align features at each stage for achieving sub-type diagnosis. Third, we leverage a progressive classifier that can progressively include additional acquired modalities (if needed) for diagnosis, to achieve the balance between diagnosis cost and diagnosis performance. We evaluate our proposed framework on large diverse public and in-home datasets (8280 subjects in total) and achieve superior performance over state-of-the-art methods.

Y. Liu and M. Liu—Equal contribution.

Supplementary Information The online version contains supplementary material available at https://doi.org/10.1007/978-3-031-78761-4_12.

Keywords: Alzheimer's disease · Multi-modality fusion · Contrastive learning · Explanation analysis · Multi-stage framework

1 Introduction

Alzheimer's disease (AD) poses a growing global health challenge, with an escalating number of subjects affected by cognitive impairment. The total healthcare costs for the treatment of AD are estimated as \$305 billion in 2023, which brings a heavy burden to society [18]. Accurate diagnosis allows effective treatment and hence alleviates AD-related burden. For AD diagnosis, key modalities encompass cognitive assessment and plasma biomarkers (recorded as tabular data), Magnetic Resonance Imaging (MRI), and Positron Emission Tomography (PET). The tabular data is relatively easier to access and thus can serve as initial screening data. For further diagnosis, MRI may be used to assess essential information regarding brain atrophy. When needed, PET (that can monitor the deposition of disease-related proteins in the brain) may be used to provide more specific evidence for AD diagnosis. Based on these modalities, deep learning (DL)-based methods usually can achieve higher diagnosis performance than the traditional ones. However, three major limitations still exist in the existing DL-based methods.

First, many DL-based studies build models for an end-to-end one-stage diagnosis using multiple modalities [4,14]. This can significantly increase diagnosis cost because a considerable number of subjects can be firmly diagnosed in early stages without additional modalities acquired in later stages. This practical challenge limits applications of these studies in clinic. **Second**, many DL-based works use only imaging data. Although there have emerged a series of works [13,14] using relatively easier-to-access tabular data for diagnosis, they simply input tabular data into the model without considering semantic information of these tabular values. Some works [10,12] try to use pre-trained language encoder to encode textualized tabular data, but they need to use detailed templates to textualize tabular data, which may inadvertently introduce more common template information, instead of specific tabular data information. **Finally**, previous methods directly classify the state of health or general AD, rather than aligning modality data with AD sub-types suggested by clinical guideline [4,6,9]. Consequently, these methods have difficulty in giving AD sub-type diagnosis results, and cannot assist in precise AD treatment in clinic.

Therefore, in this paper, we propose a progressive AD diagnosis framework (shown in Fig. 1) for cost-performance balance AD sub-type classification, which can effectively address above-mentioned limitations. We summarize our contributions as follows:

– We propose a multi-modality and multi-stage contrastive framework for AD sub-type diagnosis, which can effectively align different modalities. This alignment ensures feature(s) acquired in earlier stage(s) to also carry information from later stage(s), resulting in fewer diagnosis stages, and reducing diagnosis stages and costs.

– To better exploit clinical tabular data, we employ a text disentanglement network. This network extracts both common (from template) and specific features (from tabular data) from textualized tabular data. It minimizes similarity of specific features to let them carry different aspects of subject information. Moreover, we use a modality fusion module to fuse and enhance feature representation across different stages (if needed) for better multi-modality diagnosis performance.
– We align features in each stage with AD sub-type criteria in the clinical guideline. Moreover, based on the alignments, we develop a confidence-driven multi-stage progressive diagnosis framework. The framework automatically determines whether to give a final diagnosis based on already collected data (without further acquisition to save cost) or to further acquire more advanced modality in the next stage.
– Experimental results on large diverse public and in-home datasets (8280 subjects in total) show that our framework outperforms other representative methods significantly.

2   Methods

2.1   Notation and Problem Formulation

In alignment with real clinical procedure (often starting with easy-to-access data and later adding more difficult/expensive-to-access data if needed), our diagnostic process thus comprises *three distinct stages*, i.e., (Stage 1) using only tabular data, (Stage 2) using both tabular data and MRI data, and (Stage 3) using all the tabular data, MRI data, and PET data. Note that the early stage has a larger number of subjects for training, while the later stages always have less number of subjects for training (since most subjects have been firmly diagnosed in the early stage(s)).

For each of the stages, a classifier for AD diagnosis is required. This classifier generates two options based on the diagnosis confidence, *either* a confirmed decision *or* a demand for next-stage modality acquisition. The options rely solely on how these acquired modalities are matched with the corresponding sub-type criteria in the guideline. In the final stage, the model is compelled to generate the diagnosis decision, because there will be no additional stages. Our comprehensive framework comprises the following parts. **Multi-modality Encoder:** This part elucidates the details on multi-modality feature extraction using the corresponding encoder across stages and multi-modality fusion for diagnosis (Detailed in Sect. 2.2). **Multi-modality alignment and Clinical guideline alignment**: We show modality feature alignment among different stages and also the alignment between the fused feature with the corresponding sub-type criteria in the clinical guideline (Detailed in Sect. 2.3). **Progressive Classifier**: This part describes progressive diagnosis by referring to the diagnosis confidence (Detailed in Sect. 2.4).

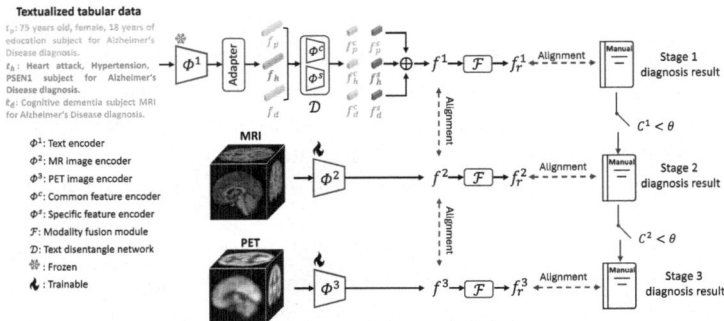

Fig. 1. We propose a progressive multi-modality and multi-stage contrastive framework for AD sub-type diagnosis. Starting from Stage 1, we acquire the features from the textualized tabular data, and align them with the corresponding sub-type criteria in the guideline. In cases where the diagnosis confidence is low, we progressively utilize MR and PET images in Stages 2 and 3, respectively. To acquire modality information of the later stage(s) without actual data acquisition, we further align the features from early stages with those from the later stages.

2.2 Multi-modality Encoder

We have three encoders Φ^1, Φ^2, and Φ^3 in three corresponding stages for processing tabular data, MR images, and PET images, respectively. At Stage k, we acquire the feature vector f^k from the corresponding modality. Specifically, in Stage 1, we use the BioMedCLIP [20] text encoder as Φ^1. For Stages 2 and 3, we train 3D image encoders Φ^2 and Φ^3 from scratch, because the BioMed-CLIP utilizes the 2D image encoder which performs poorly for 3D brain image classification.

To better accommodate Φ^1 to our task, we add an Adapter as shown in Fig. 1. We first textualize different tabular data into corresponding texts, including personal text t_p, healthy text t_h, and dementia text t_d, following the same text template (with more details are given in Sect. 3.3). Then, we separately input them into Φ^1. Each output from the Φ^1 is forwarded into the Adapter, which is a 4-layer multi-layer perceptron (MLP), and mapped to the corresponding text feature vector $f_* \in \{f_p, f_h, f_d\}$.

Previous research [10] has demonstrated that a detailed text template can make the language encoder perform better on downstream tasks. However, a highly detailed template may inadvertently introduce more generic information from the template itself rather than specific information from diverse tabular data. This can make the language encoder only acquire similar features for different textualized tabular data and thus hard to distinguish these features [15]. To address this issue, we construct a text disentanglement network $\mathcal{D}$ as formulated below.

Text Disentanglement Network. The text feature vectors (f_p, f_h, f_d) output by Φ_1 are jointly forwarded into the disentanglement network $\mathcal{D}$, as shown in Fig. 1. It aims to disentangle each feature vector $f_* \in \{f_p, f_h, f_d\}$ into a common feature vector f_*^c (from template) and a specific feature vector f_*^s (from tabular data) [2,19]. Specifically, we employ a common encoder Φ^c and a specific encoder Φ^s to map f_* into f_*^c and f_*^s, respectively. We minimize their mutual information (MI) with loss $\mathcal{L}_D^{MI} = \sum_{f_* \in \{f_p, f_h, f_d\}} MI\left(f_*^c, f_*^s\right)$. Besides, it is essential to ensure that the common features across different texts convey similar information, while the specific features highlight distinct aspects of the subject information. Therefore, we maximize the similarity of common features and minimize the similarity of specific features in all paired text features. Herein, we adopt the orthogonal loss $\mathcal{L}_D^O$ as formulated below,

$$\mathcal{L}_D^O = \frac{1}{2} \times \sum_{\substack{f_* \in \\ \{f_p, f_h, f_d\}}} \sum_{\substack{f_{**} \in \\ \{f_p, f_h, f_d\}}} -\cos\left(f_*^c, f_{**}^c\right) + \left|\cos\left(f_*^s, f_{**}^s\right)\right|. \tag{1}$$

Finally, we concatenate disentangled features of each text feature, namely $[f_*^c, f_*^s]$, $f_* \in \{f_p, f_h, f_d\}$ as each output of $\mathcal{D}$ to maximize the utilization of information from both the template and tabular data. Subsequently, all the output features from $\mathcal{D}$ are further concatenated to obtain textualized tabular data feature f^1, at Stage 1.

Multi-modality Fusion. To adaptively fuse multi-modality features from multiple stages (f^1, f^2, and f^3) for final diagnosis, we apply a transformer-based multi-modality fusion module $\mathcal{F}$ as denoted in Fig. 1. The fusion module $\mathcal{F}$ takes a transformer architecture to adaptively fuse f^1, f^2, and f^3 into a fixed-length vector for the diagnosis. Specifically, $\mathcal{F}$ first concatenates k available feature(s) at Stage k into a vector $f^{1:k} = [f^1, ... f^k], k \le 3$. Then, $f^{1:k}$ is further concatenated with learnable query Q as key and value, namely $K = V = [f^{1:k}, Q]$. Finally, the vectors Q, K, and V are used to calculate the attention ATT and feed forward block FFW [17] (an MLP with a ReLU) to finally acquire f_r^k as below:

$$f_r^k = Q + FFW\left(Q + ATT(Q, K, V)\right). \tag{2}$$

Based on the ATT and the FFW, f_r^k is mapped to the same dimension as Q. Meanwhile, since our fusion module can process various lengths of vectors, we can train our $\mathcal{F}$ using all the subjects we have.

2.3   Multi-modality Alignment and Clinical Guideline Alignment

Multi-modality Alignment. Since multiple modalities are acquired from the same subject, there often exists underlying relevance among these modalities [11]. To enhance diagnosis performance in earlier stages, we propose to let the early stage encoder(s) entail modality information of the later stage(s) without actual acquisition. Therefore, we attempt to align the modality features of all the stages. The alignment loss $\mathcal{L}_{alig}$ is formulated below:

$$\mathcal{L}_{alig} = \sum_{k=1}^{2} ||f^k - f^{k+1}||_2. \tag{3}$$

Based on the alignment of features in previous stages with the later stage ones, the features of previous stages can embody particular information that only the later stages can provide. This can improve the diagnosis performance in the previous stages, and hence save diagnosis cost.

Clinical Guideline Alignment. To achieve sub-type diagnosis with f_r^k at Stage k, we align f_r^k with the sub-type criteria in the clinical guideline. To be specific, we encode both the corresponding and non-corresponding sub-type criteria of the subject, to make f_r^k not only like its corresponding criteria Y but also different from the non-corresponding criteria $-^Y$. We define the contrastive loss $\mathcal{L}_{con}^k$ as the negative logarithm of the corresponding criteria contrastive score as expressed below:

$$\mathcal{L}_{con}^k = -\log \frac{\exp((f_r^k)\Phi^1(Y))}{\sum_{y \in -Y} \exp((f_r^k)\Phi^1(y))) + \exp((f_r^k)\Phi^1(Y))}. \tag{4}$$

During inference, we compare the contrastive score of all the AD sub-types and choose the largest one as the diagnosis decision.

2.4 Progressive Classifier

In clinical practice, it is important to consider the cost of accurate diagnosis. If the previous stage(s) can make confident diagnosis decisions already, there will be no need for later stage(s). Therefore, our goal is to minimize the total costs to the maximum extent without compromising classification performance. To this end, inspired by [21], we define diagnosis confidence C^k in each Stage k by subtracting the second highest contrastive score from the highest one in clinical guideline alignment. This subtracted value indicates how confident is the network about the diagnosis decision. Then, we set a threshold θ on C^k. As we previously mentioned in Sect. 1, there are two options for the classifier in Stage k. If diagnosis confidence C^k is larger than the threshold θ, it leads to the first option, i.e., to give the diagnosis result and stop. Otherwise, a further stage is required. Inspired by [16], we encourage the model to give confident diagnosis results in earlier stages. For each stage, we set the loss $\mathcal{L}^k$ to avoid postponement by multiplying the negative of the confidence by a penalty $\delta^k > 0$.

$$\mathcal{L}^k = \begin{cases} \mathcal{L}_{con}^k & C^k \geq \theta^k \text{ or } k = 3 \\ -C^k \delta^k & C^k < \theta^k \end{cases} \tag{5}$$

3 Experiments

3.1 Dataset

We use data from four datasets including Alzheimer's Disease Neuroimaging Initiative (ADNI) [1,7], Open Access Series of Imaging Studies (OASIS) [8], National Alzheimer's Coordinating Center (NACC) [3], and an in-house dataset from Huashan Hospital. We follow the IWG-2 diagnosis guideline [5] and group the subjects into four sub-types, including typical, atypical, pre-clinical AD, and normal control. The corresponding sub-type criteria are detailed in the supplementary material. We have totally 8280 subjects with tabular data, where 4842 subjects have both tabular and MRI data and 2336 subjects have all the tabular, MRI, and PET data. Datasets are randomly divided into 5 equal folds. 3 folds are used for training and the remaining 2 folds for validation and testing. The model is tested on subjects with full modalities. We do not divide the dataset with overlapping subjects to avoid the leakage. Note that although we use dementia information, this will not cause the data leakage as IWG-2 [5] considers both dementia information and in-vivo evidence for AD diagnosis instead of dementia information only.

3.2   Implementation Details

To ensure a fair comparison, we use 3D CNN as image encoders (as detailed in supplementary material) for all comparison methods. We use the SGD optimizer with a learning rate of 10^{-4}. We train our model for 100 epochs. We choose the MSE as the $\mathcal{L}_{alig}$. We set θ as 0.3, and weights for all the loss functions are 1. δ^1 and δ^2 are set to be 1 and 1.5, respectively. For text input t_p, we consider age, education, and gender. For t_h, we consider healthy information, including heart attack, hypertension, stroke, alcohol abuse, psychiatric disorder, and the blood test. For t_d, we employ dementia level. As some subjects may lack specific tabular information, we only input available textualized tabular data to the text encoder.

3.3   Results and Discussions

We employ six distinct evaluation metrics to assess both diagnostic performance and cost-effectiveness for subjects with full modalities (837 subjects). The first four metrics are for sub-type diagnosis and the other two are for cost-effectiveness. For estimating costs, we use the mean stage number as a proxy. Finally, we compute the ratio of AUC to the estimated cost to measure the cost-effectiveness.

Table 1. Quantitative comparison with other methods on our test dataset.

Models	Acc	Spe	Sens	AUC	Cost	AUC/Cost ratio
CoCoOp	74.1	83.3	74.6	72.9	2.80	26.03
Xplainer	75.3	85.6	73.2	75.5	2.53	29.84
Ours w/o progressive classifier	80.8	87.1	80.6	81.7	3	27.23
Ours w/o multi-modality alignment	76.5	86.8	76.6	75.3	2.45	30.73
Ours w/o $\mathcal{D}$	74.5	83.6	71.6	73.8	2.42	30.50
Ours w/o $\mathcal{F}$	75.8	84.9	76.2	73.7	2.52	29.24
Ours	78.2	88.3	77.8	77.4	2.21	**35.02**

From Table 1, we can observe that our framework demonstrates a superior classification performance over the representative methods (CoCoOp [22] and Xplainer [12]), which are also built on multi-stage and multi-modality fusion. These two comparison methods use different text encoders, but the same image encoder as ours. Our performance improvement mainly arises from two reasons. *First*, we use text disentanglement network $\mathcal{D}$ to extract more information from different textualized tabular data. *Second*, we further propose multi-modality alignment to let early-stage features entail later-stage information, allowing the model to have improved diagnosis performance in the early stages for reducing acquisition cost. The ablation studies on w/o $\mathcal{D}$ and multi-modality alignment further validate our conclusion. Besides, we also show the effectiveness of the progressive classifier. We can see that, although there exists a certain performance drop in diagnosis accuracy compared to the one without using the progressive classifier, it is expected, since in earlier stages, we use fewer modalities for diagnosis. However, the use of the progressive classifier achieves a substantial reduction in diagnosis cost, and also improves the AUC/Cost ratio significantly, implying that our model has great potential for practical applications in clinical routine.

Text Templates Design. In our evaluation, we test three types of text templates as outlined below (by using age information in t_p as an example). The first template (template 1) solely focuses on basic tabular information of the subject. For templates 2 and 3, we introduce more detailed descriptions. The corresponding performance is depicted in Table 2. We can observe that, employing a more detailed text template can enhance performance. Hence, we choose to utilize template 3 in our method. Besides, we can see that the more detailed the template is, the more contribution $\mathcal{D}$ can make to diagnosis performance. This demonstrates that $\mathcal{D}$ can more effectively extract features from the detailed text templates.

- **Template 1:** {75 years old}
- **Template 2:** {75 years old} subject
- **Template 3:** {75 years old} subject for Alzheimer's Disease diagnosis

Table 2. Evaluation on different text templates and text disentanglement network $\mathcal{D}$.

| Template | w/ $\mathcal{D}$ | | | | | | w/o $\mathcal{D}$ | | | | | |
	Acc	Spe	Sens	AUC	Cost	AUC/Cost ratio	Acc	Spe	Sens	AUC	Cost	AUC/Cost ratio
Template 1	70.7	78.3	71.9	71.5	2.39	29.91	69.8	77.7	70.5	71.2	2.46	28.94
Template 2	73.1	80.9	73.8	73.5	2.35	31.27	71.2	78.7	70.9	71.5	2.48	28.83
Template 3	78.2	88.3	77.8	77.4	2.21	35.02	74.5	83.6	71.6	73.8	2.42	30.50

Effect of Threshold on Confidence. We delve deeper into exploring the impact of using varying confidence thresholds in the progressive classifier. A relatively lower confidence threshold encourages the model to make decisions in earlier stage(s) and substantially decreases acquisition costs. However, it may also lead to incorrect diagnosis results that cannot be corrected by subsequent stage(s). On the contrary, opting for a higher confidence threshold can lead to more accurate diagnosis results but cause increased diagnosis costs. In Table 3, we present performance and cost ratio using different confidence thresholds. We set the same threshold value in stages 1 and 2. As we can see, as the threshold increases, diagnosis performance is also improved, but the improvement ratio slows down. With regard to the AUC/Cost ratio, it increases first and then decreases dramatically. This effect is reasonable since, when the threshold becomes too large, most of the subjects are required to finish later stages to obtain diagnosis decisions. This will benefit diagnosis correctness, but with the sacrifice of the AUC/cost ratio since many subjects can be firmly diagnosed in early stages without going to the next expensive stage(s).

Table 3. Evaluation on different confidence thresholds.

Threshold	Acc	Spe	Sens	AUC	Cost	AUC/Cost ratio
0.1	71.8	70.9	72.3	72.2	2.09	34.54
0.3	78.2	88.3	77.8	77.4	2.21	35.02
0.5	78.9	87.7	79.3	79.6	2.63	20.26

4   Conclusion

In this work, we present a multi-modal and multi-stage framework for cost-effective AD sub-type diagnosis. Different from the existing works, our framework is to progressively give diagnosis results from a single modality to multiple modalities. Specifically, in the first stage, we utilize only tabular data. To effectively extract textualized tabular information, we propose a text disentanglement network to disentangle common and specific features of different tabular data. If diagnosis confidence of the first stage is lower than a pre-defined threshold, the respective subject requires later stages, which integrates the MRI and PET data. To effectively fuse multiple modalities, we propose a transformer-based multi-modality fusion module. To improve diagnosis performance of early stages and reduce overall diagnosis costs, we propose to align features of different stages to allow early stages to contain particular information only available in later stage(s). Furthermore, we match modality features and the AD diagnosis guideline to achieve improved sub-type diagnosis. Extensive experiments demonstrate the superiority of our framework, showing promise in real AD diagnosis.

Acknowledgments. This work was supported in part by National Natural Science Foundation of China (grant numbers U23A20295, 62131015), the STI 2030Major Projects (No. 2022ZD0209000), Shanghai Municipal Central Guided Local Science and Technology Development Fund (grant number YDZX20233100001001), and The Key R&D Program of Guangdong Province, China (grant numbers 2023B0303040001, 2021B0101420006).

Disclosure of Interests. The authors have no competing interests to declare that are relevant to the content of this article.

References

1. Aisen, P.S., Petersen, R.C., Donohue, M., Weiner, M.W.: Alzheimer's Disease Neuroimaging Initiative 2 Clinical Core: Progress and Plans (2015)
2. Balasubramanian, V., Kobyzev, I., Bahuleyan, H., Shapiro, I., Vechtomova, O.: Polarized-VAE: Proximity based disentangled representation learning for text generation. arXiv preprint arXiv:2004.10809 (2020)
3. Beekly, D.L., et al.: The national Alzheimer's coordinating center (NACC) database: the uniform data set. Alzheimer Dis. Assoc. Disord. **21**(3), 249–258 (2007)
4. Cheng, D., Liu, M.: CNNs based multi-modality classification for ad diagnosis. In: 2017 10th International Congress on Image and Signal Processing, Biomedical Engineering and Informatics (CISP-BMEI), pp. 1–5. IEEE (2017)
5. Dubois, B., et al.: Advancing research diagnostic criteria for Alzheimer's disease: the IWG-2 criteria. Lancet Neurol. **13**(6), 614–629 (2014)
6. Huang, Y., Xu, J., Zhou, Y., Tong, T., Zhuang, X.: Alzheimer's disease neuroimaging initiative (ADNI). Diagnosis of Alzheimer's disease via multi-modality 3D convolutional neural network. Front. Neurosci. **13**, 509 (2019)
7. Jack, C.R.: Magnetic Resonance Imaging in Alzheimer's Disease Neuroimaging Initiative 2 (2015)
8. LaMontagne, P.J., et al.: OASIS-3: longitudinal neuroimaging, clinical, and cognitive dataset for normal aging and Alzheimer disease. MedRxiv, pp. 2019–12 (2019)
9. Liu, F., Wee, C.-Y., Chen, H., Shen, D.: Inter-modality relationship constrained multi-modality multi-task feature selection for Alzheimer's disease and mild cognitive impairment identification. Neuroimage **84**, 466–475 (2014)
10. Liu, J., et al.: Clip-driven universal model for organ segmentation and tumor detection. In: Proceedings of the IEEE/CVF International Conference on Computer Vision, pp. 21152–21164 (2023)
11. Pan, Y., Liu, M., Lian, C., Zhou, T., Xia, Y., Shen, D.: Synthesizing missing pet from MRI with cycle-consistent generative adversarial networks for Alzheimer's disease diagnosis. In: Medical Image Computing and Computer Assisted Intervention–MICCAI 2018: 21st International Conference, Granada, Spain, September 16-20, 2018, Proceedings, Part III 11, pp. 455–463. Springer (2018)
12. Pellegrini, C., Keicher, M., Özsoy, E., Jiraskova, P., Braren, R., Navab, N.: Xplainer: From x-ray observations to explainable zero-shot diagnosis. arXiv preprint arXiv:2303.13391 (2023)
13. Pölsterl, S., Wolf, T.N., Wachinger, C.: Combining 3D image and tabular data via the dynamic affine feature map transform. In: de Bruijne, M., et al. (eds.) MICCAI 2021. LNCS, vol. 12905, pp. 688–698. Springer, Cham (2021). https://doi.org/10.1007/978-3-030-87240-3_66

14. Qiu, S., et al.: Multimodal deep learning for Alzheimer's disease dementia assessment. Nat. Commun. **13**(1), 3404 (2022)
15. Seibold, C., Reiß, S., Sarfraz, M.S., Stiefelhagen, R., Kleesiek, J.: Breaking with fixed set pathology recognition through report-guided contrastive training. In: International Conference on Medical Image Computing and Computer-Assisted Intervention, pp. 690–700. Springer (2022)
16. Trapeznikov, K., Saligrama, V., Castañón, D.: Multi-stage classifier design. In: Asian Conference on Machine Learning, pp. 459–474. PMLR (2012)
17. Vaswani, A., et al.: Attention is all you need. In: Advances in Neural Information Processing Systems, vol. 30 (2017)
18. Wong, W.: Economic burden of Alzheimer disease and managed care considerations. Am. J. Manag. Care **26**(8 Suppl), S177–S183 (2020)
19. Zhang, J., Hong, H., Zhang, Y., Wan, Y., Liu, Y., Sui, Y.: Disentangled code representation learning for multiple programming languages. In: Findings of the Association for Computational Linguistics: ACL-IJCNLP 2021 (2021)
20. Zhang, S., et al.: Large-scale domain-specific pretraining for biomedical vision-language processing. arXiv preprint arXiv:2303.00915 (2023)
21. Zheng, Z., Teng, S., Naiqi, W., Teng, L., Zhang, W., Fei, L.: Selected confidence sample labeling for domain adaptation. Neurocomputing **555**, 126624 (2023)
22. Zhou, K., Yang, J., Loy, C.C., Liu, Z.: Conditional prompt learning for vision-language models. In: Proceedings of the IEEE/CVF Conference on Computer Vision and Pattern Recognition, pp. 16816–16825 (2022)

Surface-Based Parcellation and Vertex-wise Analysis of Ultra High-resolution *ex vivo* 7 tesla MRI in Alzheimer's disease and related dementias

Pulkit Khandelwal[1]([✉]), Michael Tran Duong[1], Lisa Levorse[1],
Constanza Fuentes[2], Amanda E. Denning[1], Winifred Trotman[1],
Ranjit Ittyerah[1], Alejandra Bahena[1], Theresa Schuck[1], Marianna Gabrielyan[1],
Karthik Prabhakaran[1], Daniel T. Ohm[1], Gabor Mizsei[1], John Robinson[1],
Monica Muñoz[2], John A. Detre[1], Edward B. Lee[1], David J. Irwin[1],
Corey McMillan[1], M. Dylan Tisdall[1], Sandhitsu R. Das[1], David A. Wolk[1],
and Paul A. Yushkevich[1]

[1] University of Pennsylvania, Philadelphia, USA
pulks@seas.upenn.edu
[2] University of Castilla-La Mancha, Ciudad Real, Spain

Abstract. Magnetic resonance imaging (MRI) is the standard modality to understand human brain structure and function *in vivo* (antemortem). Decades of research in human neuroimaging has led to the widespread development of methods and tools to provide automated volume-based segmentations and surface-based parcellations which help localize brain functions to specialized anatomical regions. Recently *ex vivo* (postmortem) imaging of the brain has opened-up avenues to study brain structure at sub-millimeter ultra high-resolution revealing details not possible to observe with *in vivo* MRI. Unfortunately, there has been limited methodological development in *ex vivo* MRI primarily due to lack of datasets and limited centers with such imaging resources. Therefore, in this work, we present one-of-its-kind dataset of **82** *ex vivo* **T2w whole-brain hemispheres MRI at 0.3 mm**^{3} resolution spanning Alzheimer's disease and related dementias. We adapted and **developed a fast and easy-to-use** automated surface-based pipeline to parcellate, for the first time, ultra high-resolution *ex vivo* brain tissue at the native subject-space resolution using the Desikan-Killiany-Tourville (DKT) brain atlas. This allows us to perform vertex-wise analysis in the template space and thereby link morphometry measures with pathology measurements derived from histology. We open-source our code, docker container and Jupyter notebooks for ready-to-use out-of-the-box set of tools and command line options to advance *ex vivo* MRI clinical brain imaging research at the project webpage.

Keywords: exvivo MRI · 7 tesla · surface parcellation · segmentation

© The Author(s), under exclusive license to Springer Nature Switzerland AG 2025
D. R. Bathula et al. (Eds.): MLCN 2024, LNCS 15266, pp. 134–144, 2025.
https://doi.org/10.1007/978-3-031-78761-4_13

1 Introduction

Alzheimer's disease and related dementias (ADRD) are heterogeneous, with multiple neuropathological processes jointly contributing to neurodegeneration [34]. Currently, co-pathologies are not reliably detected with *in vivo* imaging, which makes it difficult for clinicians to accurately diagnose the cause of cognitive impairment in individual patients and to identify those most likely to benefit from treatment with emergent drugs that target AD pathology. The efficacy of *in vivo* biomarkers can be enhanced by combined analysis of regional measures of neurodegeneration, such as cortical thickness, derived from brain MRI (*ex vivo* or *in vivo*) and pathological measures obtained at autopsy. Such analysis can help identify distinct spatial patterns of disease progression linked to AD-specific pathologies (i.e. β-amyloid plaques and phosphorylated tau protein tangles) and common co-pathologies. While such studies are performed using either *in vivo* or *ex vivo* MRI, the latter offers advantages in temporal proximity to pathological markers and in spatial resolution. The ability to scan *ex vivo* brain tissue for many hours without motion artifacts makes it possible to achieve ultra-high spatial resolution at which intricate neuroanatomical details of the brain can be revealed [1,2,6,10,20,32,36,37]. **However, lack of automated tools for processing ultra high-resolution *ex vivo* MRI is a significant barrier to study structure-pathology association in ADRD.**

Decades of neuroimaging research has yielded advanced computational frameworks for automated analysis of *in vivo* brain MRI, and tools such as FreeSurfer [13] and Statistical Parametric Mapping (SPM) [5], FSL [29] have been applied in a plethora of ADRD neuroimaging studies. However, these tools cannot be directly used on ultra-high resolution *ex vivo* MRI and there is very limited work on developing *ex vivo* MRI analysis methods for widespread use, which remains a challenge primarily due to the greater heterogeneity in scanning protocols of *ex vivo* MRI and increased complexity and imaging artifacts than *in vivo* MRI. Recent developments for *ex vivo* MRI include deep learning methods for high resolution cytoarchitectonic mapping in 2D histology [3], atlas-based segmentation of MTL and the thalamus [20,21], and whole-brain hemispheres analysis [27,28]. Recently, deep learning-based methods were developed to segment the granular cortical layers [25], and gray matter (GM), subcortical structures, white matter (WM) and its hyperintensities (WMH) in 7T *ex vivo* MRI. NextBrain [8] provides brain segmentations using histology brain atlas.

These previous methods have several limitations as they are specialized for specific parts of the brain or specific *ex vivo* datasets mainly limited to a healthy (unremarkable) brain tissue specimens. The methods were evaluated on very small datasets of only 1 [12], 5 [8], and 17 [25] *ex vivo* brain tissue for unremarkable specimens or at low-resolution [28]. Crucially, to our knowledge, **the feasibility of using existing *ex vivo* MRI analysis pipelines to perform large-scale structure-pathology association studies in ultra-high resolution *ex vivo* MRI has not been demonstrated.**

Contributions. In this study, we perform structure-pathology association analysis in a large dataset of ultra high-resolution $0.3\,\text{mm}^3$ T2w 7T MRI scans of whole brain hemispheres from 82 brain donors with ADRD diagnoses, a first study of such scale conducted at this resolution. We present a new computational pipeline that performs automated segmentation and **whole-hemisphere FreeSurfer DKT atlas [11] parcellation of the cortex in native subject space** at sub-millimeter 0.3 mm ^{3} resolution, a **first large-scale surface-based scheme for** *ex vivo* **whole-hemispheres analysis in diseased population.** We achieve this by adapting the surface-based modeling steps in FreeSurfer with an initial subject-space topology-corrected WM segmentation derived from a deep learning-based segmentation model as developed in [24]. We evaluate the framework by correlating cortical thickness with neuropathological markers implicated in AD (measures of p-tau, neuronal loss; global amyloid-β, Braak staging, and CERAD ratings) and perform vertex-wise generalized linear modeling.

2   Materials and Methods

Dataset. We acquire and analyze a dataset of **82 ex vivo whole hemisphere 0.3 mm^3 7T MRI** (See Fig. 2) with left: 42 and right: 40 hemispheres, postmortem interval (PMI) of 18.48 ± 13.60 h and fixation time of 256.70 ± 280.48 days. Patients were evaluated at the Penn Frontotemporal Degeneration Center (FTDC) or Alzheimer's Disease Research Center (ADRC) and followed to autopsy at the Penn Center for Neurodegenerative Disease Research (CNDR) center [24]. The cohort included patients with Alzheimer's Disease or related dementias (ADRD) [26] spectrum, such as Lewy body disease, limbic-predominant age-related TDP-43 encephalopathy, corticobasal degeneration, primary age-related tauopathy comprising of 41 female (age: 76.97 ± 9.70 years) and 41 male (age: 76.48 ± 11.67 years). Human brain specimens were obtained in accordance with local laws and regulations, and includes informed consent from next of kin at time of death. After autopsy, one hemisphere was fixed in formalin for at least 4 weeks and then imaged at T2-w 7T using a 3D-encoded T2 SPACE sequence with 0.3 mm^3 isotropic resolution, 3 s repetition time (TR), echo time (TE) 383 ms, turbo factor 188, echo train duration 951 ms, bandwidth 348 Hz/px with 2–3 hours per scan.

The non-imaged hemisphere (contralateral tissue) underwent histological processing for neuropathological examination [24]. Tissue blocks were embedded in paraffin, sectioned at 6 μm thickness, and underwent immunohistochemistry. In the medial temporal lobe (MTL) region, semi-quantitative severity ratings of p-tau pathology and neuronal loss were assigned by expert neuropathologists on a scale of 0–3 i.e. '0: None/Rare', '1: Mild', '2: Moderate' or '3: Severe'. Standardized global AD progression ratings were also derived, including Thal stage (global amyloid-β), Braak (global p-tau), and CERAD stage (neuritic plaques) [4].

Methodological Pipeline. We present a fast and reliable anatomical parcellation and cortical thickness quantification framework that combines deep learning-based methods developed in [24] with classical methods for topology correction [33] and surface-based cortical modeling, inflation, registration, and parcellation in FreeSurfer [13, 14] for whole-hemisphere *ex vivo* brain MRI. Our framework processes one hemisphere in around 60 min. Rather than reinventing the wheel, our approach focuses on combining the strengths of these existing tools and making modifications that enable them to scale to ultra-high resolution *ex vivo* MRI. See Fig. 1 for the schematic pipeline.

Volume-Based Segmentation: First, we train a deep learning segmentation model nnU-Net [23] as explained in [24] to obtain whole-hemisphere volume-based segmentations of 8 regions namely: cortical gray matter (GM), white matter (WM), white matter hyperintensities (WMH), ventricles and four subcortical structures: caudate, putamen, globus pallidus and thalamus (Fig. 1 column A). The model was trained on Nvidia Quadro RTX 5000 GPU using manual-labeled images and extensively validated, as detailed in the work in our previous work [24]. The inference time for each hemisphere is around 15 min on a CPU. We merge the WM, WMH, ventricles and subcortical labels in a single label and employ the post-hoc topology correction method [7, 16] based on fast-marching algorithm implemented in 'Nighres/CRUISE' [18] to solve the *buried sulcus* problem, i.e., clearly separating the adjoining gyri in the opposite banks of a buried sulcus. This correction step takes around 5 min on a CPU.

Surface-Based Parcellation: Next, we modify the surface-based scheme in FreeSurfer to obtain DKT atlas-based surface parcellations in *ex vivo* native subject space. Of note is that this step does not require the raw MRI intensity image and is purely based on the voxel-level segmentation of cortical GM from above. First, a 'filled' segmentation with WM, WMH, ventricles and the four subcortical structures as the foreground; and the cortical GM along with the CSF as the background, is created from the initial volume-based segmentation. This 'filled' segmentation (Fig. 1 column B) is then tessellated and corrected for any topological errors such as holes and handles based on [33], termed as the 'WM surface'. The 'WM surface' is smoothed and inflated to a sphere (Fig. 1 column C) which is then registered to the spherical representation of the Buckner40 spherical atlas [22] provided in FreeSurfer using the curvature map of the WM surface (Fig. 1 column D). The labels of the DKT-atlas is then warped from the spherical-atlas to the spherical-representation of the *ex vivo* subject. Next, the 'WM surface' is deformed and re-positioned to find the GM-CSF pial surface using the method described in [13, 14] which employs three energy functionals: spring-like term, decomposed into tangential and normal components, to smooth and for regularization of the surface. The third term in the energy functional is dependent on the gradient of the intensity image. This term is the classic stopping criteria (inverse of the image gradient) used in deformable models based on level sets and fast marching methods. FreeSurfer uses raw *in vivo* T1w (or where applicable T2w) MRI intensity values for this energy term. For the current study, the T2w *ex vivo* MRI has a varying intensity profile within the GM which causes the deformable surface to prematurely stop in the GM cortical ribbon

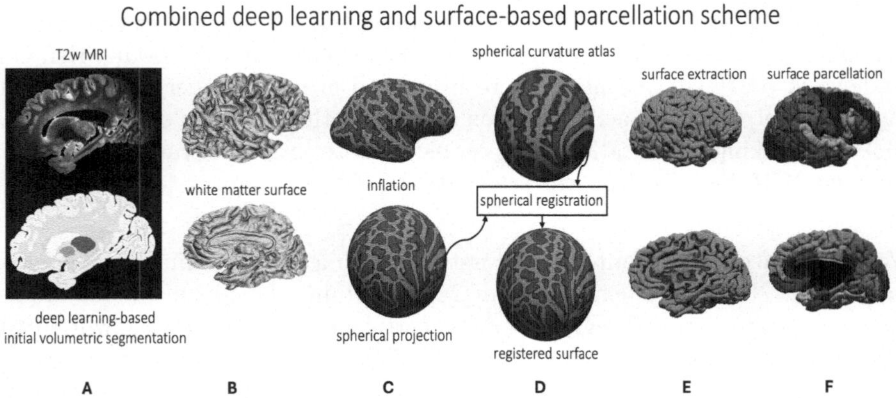

Combined deep learning and surface-based parcellation scheme

Fig. 1. Schematic of the developed pipeline based on deep learning volumetric segmentations and surface-based modeling for parcellations of *ex vivo* **whole hemisphere 0.3 mm^3 7T MRI**. The sequential steps follow A-F as described in Sect. 2.

and does not reach the pial surface. Therefore, we use the segmentation derived from the above volume-based step in place of the T2w *ex vivo* MRI intensity image. This has a clear advantage as the intensity is constant in the GM cortical ribbon with a zero gradient everywhere. The evolving surface then stops at the pial surface as expected (Fig. 1 column E). Finally, the DKT-atlas labels are projected onto the pial surface and the parcellations (Fig. 1 column F) are mapped to the voxel-space to get the region-wise segmentation in the volume. This surface-based scheme takes 40 min on a CPU with 64 threads.

3   Experiments and Results

Segmentation and Parcellation. We deployed the deep-learning model to produce the initial segmentation of GM, WM, WMH, ventricles and the subcortical structures. The deep learning-based pipeline was extensively validated in a previous work [24] as explained in Sect. 2. The surface-based scheme was then used to jointly obtain the cortical parcellations and volumetric segmentations as shown in Fig. 2 in the native subject space. Due to the immense manual labelling efforts involved in *ex vivo* MRI, in the order of 3–4 weeks per subject, the reliability of these parcellations are evaluated using the following two correlation studies. Also note that all the left hemispheres were flipped to right so that all the hemispheres are in the same orientation for parcellation and downstream analyses.

Region-Based Cortical Thickness vs Neuropathology Correlations. Figure 3 shows the Spearman's ρ-value between the mean cortical thickness (mm) in each brain region (computed in subject space at native resolution) and five pathology measures: global ratings of amyloid-β, Braak staging, CERAD, and

Volumetric segmentation and surface parcellations of ex vivo 7T T2w MRI

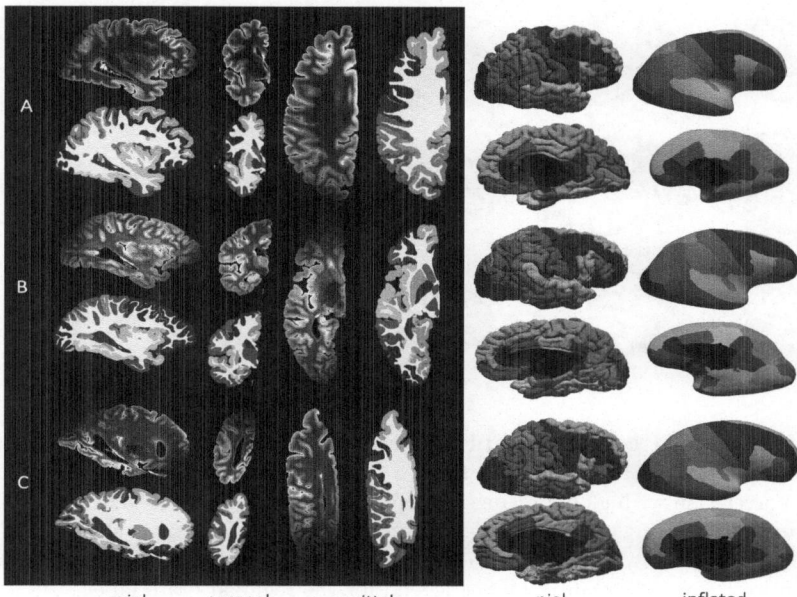

axial   coronal   sagittal   pial   inflated

Fig. 2. Ex vivo MRI segmentations and parcellations. Axial, coronal and sagittal viewing planes of *ex vivo* MRI at $0.3\,\mathrm{mm}^3$ resolution for three subjects (A, B and C) with corresponding DKT volumetric segmentations and surface-based parcellations on pial and inflated surfaces for the medial and lateral views in native subject space resolution. Our method is able to correctly delineate the brain even in regions where the MR signal contrast is low in the anterior and the posterior brain MRI due to artifacts in acquisition protocol. Legend: See Fig. 3.

regional ratings of neuronal loss and p-tau pathology in the MTL, the region first implicated in AD. The analysis were covaried for age, sex and postmortem interval (PMI) and corrected for multiple comparisons using Bonferroni method. Significant negative correlation were found in entorhinal, parahippocampal, medial-orbitofrontal, temporal pole, inferior temporal and parietal lobes which are consistent with literature on progressive loss of cortical gray matter in AD [35].

Surface-Based Vertex-Wise Cortical Thickness vs Neuropathology Correlations. The thickness maps for each individual subject were warped to the Buckner40 template-space for vertex-wise correlation analysis between cortical thickness (mm) and the five neuropathological ratings. The parcellations and thickness maps of the three subjects (Fig. 1) in the Buckner40 *fsaverage* space are shown in Supplemental Fig. 1. Vertex-wise correlation between thickness and the neuropathology ratings was performed by fitting a GLM at each vertex across the entire cohort of 82 subjects with age, sex and PMI as the covariates and corrected for multiple comparisons using family-wise error rate

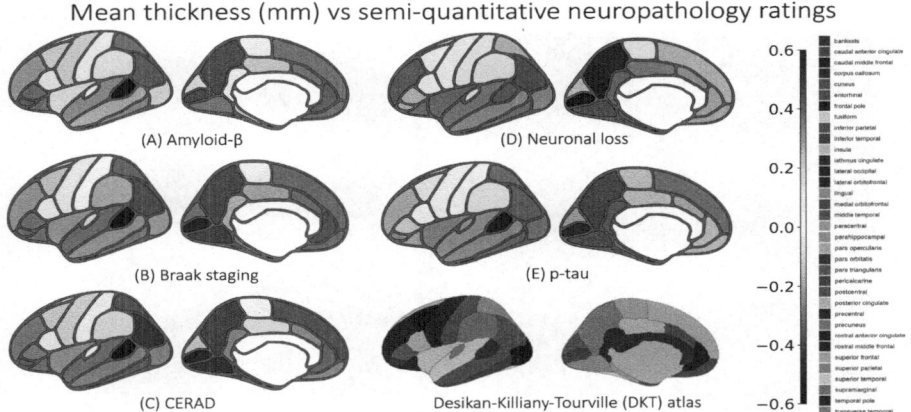

Fig. 3. Spearman's correlation plots between mean ROI thickness (mm) and neuropathological ratings in native subject-space. We observe significant negative correlation with global ratings of amyloid-β, Braak staging, CERAD, and the semi-quantitative ratings of the medial temporal lobe (MTL) neuronal loss and tau pathology. All the analysis were covaried for age, sex and postmortem interval (PMI) for the entire cohort of 82 subjects. See legend for the regional brain labels.

(FWER) correction. Figure 4 shows the t-statistics map on the pial surface with the clusters outlined in white indicating regions where the significant strongest associations were observed (p<0.05) surviving FDR correction. We observe that the strongest correlations were observed in MTL, the region associated with AD.

Comparison with Other Methods and Parcellations with Other Atlases. Additionally, we also compared our method with two other segmentation methods SynthSeg [38] (designed for contrast-agnostic MRI) and NextBrain [8] (designed specifically for ex vivo MRI). See Supplemental Fig. 2 for the qualitative results. Both SynthSeg and NextBrain fail to provide meaningful segmentations of *ex vivo* MRI. SynthSeg (operating at lower $1\,\mathrm{mm}^3$) roughly delineates major brain regions but fails at segmenting the finer details in the cortical ribbon and bleeds into CSF. Whereas, NextBrain incorrectly segments the entire cortical ribbon as it does not demarcate the different sulci and gyri. In contrast, our method clearly demarcates the brain regions in native $0.3\,\mathrm{mm}^3$ subject-space high resolution using the DKT-atlas. After visual quality control of segmentations produced by both SynthSeg and NextBrain, we concluded that a quantitative analysis similar to our proposed method and experiments is not feasible on the segmentations produced by these methods as depicted in Supplemental Fig. 2. Separately, in addition to the DKT atlas, we also provide parcellations using the Schaefer [39], Glasser [40] and the Von Economo-Koskinos [41] atlases in native subject-space resolutions for *ex vivo* brain hemisphere as shown in Supplemental Fig. 3.

4   Discussion

Using both recent advances in deep learning based segmentation models and classic deformable and spherical registration methods, we developed a pipeline that enables surface-based modeling and group-wise registration of the cortical gray matter in ultra high-resolution *ex vivo* MRI. In addition to its higher resolution, *ex vivo* MRI poses multiple challenges for computational analysis, including deformation during brain removal and fixation (particularly collapse of the ventricles), and distinct artifacts, such as intensity inhomogeneity both across the whole image domain and across cortical layers. Hence our ability to successfully combine nnU-Net, Nighres/CRUISE and FreeSurfer pipelines to achieve the same type of parcellation as is ubiquitous in *in vivo* MRI studies is a non-trivial accomplishment. Crucially, the Nighres/CRUISE and FreeSurfer components used in our pipeline operate only on segmentation maps, rather than MRI intensity. Thus, to generalize our approach to other scanners and ultra-high resolution MRI protocols, one only needs to focus on the initial segmentation aspect.

A key contribution of this work is that it performs, for the first time, a surface-based structure-pathology correlation analysis using ultra-high resolution *ex vivo* MRI in a large cohort of brain donors with ADRD diagnoses. By doing so, it demonstrates the analysis approach that is ubiquitous in *in vivo* brain MRI research, and has been previously adopted to $1mm^3$ resolution *ex vivo* MRI by some groups [27,28], can also be applied to *ex vivo* studies at ultra-high resolution. To demonstrate analysis modes common in *in vivo* research, we performed both region of interest (ROI)-level correlation analysis and vertex-wise template-space analysis to study patterns of neurodegeneration within the

Generalized linear model (GLM) for vertex-wise thickness (mm) vs neuropathology ratings

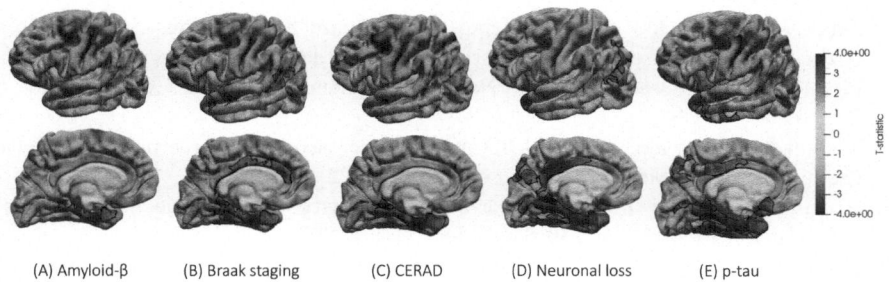

(A) Amyloid-β   (B) Braak staging   (C) CERAD   (D) Neuronal loss   (E) p-tau

Fig. 4. Template-space vertex-wise morphometry-pathology correlations. Vertex-wise group analysis was performed to fit a generalized linear model (GLM). Shown are the statistical map (t-statistics) of the correlation between cortical thickness (mm) and with global ratings of amyloid-β, Braak staging, CERAD, and semi-quantitative ratings of the medial temporal lobe (MTL) neuronal loss and tau pathology, with age, sex and postmortem interval (PMI) as covariates across all 82 subjects. The clusters outlined in black indicate regions significant correlations (p<0.05) were observed after FWER correction for multiple comparisons.

AD continuum. Prior works [15,31] shows how amyloid-β and p-tau etc. have differential influences on cortical atrophy. Tau pathology is concurrent with neuronal loss in AD leading to cortical atrophy [17]. Negative correlations between MTL p-tau and neuronal loss with cortical thickness were found to be significant in the entorhinal cortex and the MTL, regions first implicated in AD [9,30] both in *in vivo* and *ex vivo* studies. Crucially, the ability to map data from subject space to the DKT-template space based on surface correspondence is not limited to cortical thickness measures, and in the future, our pipeline can be used to analyze features that take advantage of ultra-high resolution *ex vivo* MRI, such as mapping iron and myelin in cortex [36].

Conclusion. We developed a fast and easy-to-use pipeline to parcellate the whole-brain hemispheres into the FreeSurfer DKT-atlas defined regions at the native subject space sub-millimeter $0.3\,\text{mm}^3$ isotropic resolution, and applied it to a large-scale dataset of ultra high-resolution *ex vivo* 7T structural MRI. We demonstrated the feasibility and utility of this approach by performing a structure-pathology association study that is the first of its kind at this level of *ex vivo* MRI resolution and one of the largest ever conducted at any *ex vivo* MRI resolution. In the future, we will perform an exhaustive quantitative study by delineating the brain regions with different anatomical atlases based on a surface-based population template derived from the *ex vivo* hemispheres. This will enable us to understand the region-wise local intensity profiles to study the structural relationships with the underlying histology-derived markers. We believe that the present study will enable scientific advancements in *ex vivo* imaging and thereby inform better *in vivo* computational biomarkers.

References

1. Adler, D.H., et al.: Characterizing the human hippocampus in aging and Alzheimer's disease using a computational atlas derived from ex vivo MRI and histology. In: Proceedings of the National Academy of Sciences vol. 115, no. 16 (2018)
2. Alkemade, A., et al.: A unified 3D map of microscopic architecture and MRI of the human brain. Sci. Adv. **8**(17), eabj7892 (2022)
3. Amunts, K., et al.: Julich-brain: a 3D probabilistic atlas of the human brain's cytoarchitecture. Science **369**(6506), 988–992 (2020)
4. Mirra, S.S., et al.: The consortium to establish a registry for Alzheimer's disease (CERAD): Part II. Standardization of the neuropathologic assessment of Alzheimer's disease. Neurology **41**(4), 479–479 (1991)
5. Ashburner, J.: Computational anatomy with the SPM software. Magn. Reson. Imaging **27**(8), 1163–1174 (2009)
6. Augustinack, J.C., van der Kouwe, A.J., Fischl, B.: Medial temporal cortices in ex vivo magnetic resonance imaging. J. Comp. Neurol. **521**(18), 4177–88 (2013)
7. Bazin, P.L., Pham, D.L.: Topology correction of segmented medical images using a fast marching algorithm. Comput. Methods Programs Biomed. **88**(2), 182–190 (2007)

8. Casamitjana, A., et al.: A next-generation, histological atlas of the human brain and its application to automated brain MRI segmentation. bioRxiv 2024 (2024)

9. Das, S.R., et al.: In vivo measures of tau burden are associated with atrophy in early Braak stage medial temporal lobe regions in amyloid-negative individuals. Alzheimer's Dement. **15**(10), 1286–1295 (2019)

10. DeKraker, J., et al.: Surface-based hippocampal subfield segmentation. Trends Neurosci. **44**(11), 856–863 (2021)

11. Desikan, R.S., et al.: An automated labeling system for subdividing the human cerebral cortex on MRI scans into gyral based regions of interest. Neuroimage **31**(3), 968–980 (2006)

12. Edlow, B.L., et al.: 7 Tesla MRI of the ex vivo human brain at 100 micron resolution. Sci. Data **6**(1), 244 (2019)

13. Fischl, B., et al.: Cortical surface-based analysis: II: inflation, flattening, and a surface-based coordinate system. Neuroimage **9**(2), 195–207 (1999)

14. Fischl, B., et al.: Cortical surface-based analysis: II: inflation, flattening, and a surface-based coordinate system. Neuroimage **9**(2), 195–207 (1999)

15. Frigerio, I., et al.: Amyloid-β, p-tau and reactive microglia are pathological correlates of MRI cortical atrophy in Alzheimer's disease. Brain Commun. **3**(4), fcab281 (2021)

16. Han, X., et al.: CRUISE: cortical reconstruction using implicit surface evolution. NeuroImage **23**(3), 997–1012 (2004)

17. Harrison, T.M., et al.: Distinct effects of beta-amyloid and tau on cortical thickness in cognitively healthy older adults. Alzheimer's Dementia **17**, 1085–1096 (2021)

18. Huntenburg, J.M., et al.: Nighres: processing tools for high-resolution neuroimaging. GigaScience. **7**(7), giy082 (2018)

19. Hyman, B.T., et al.: National Institute on Aging-Alzheimer's Association guidelines for the neuropathologic assessment of Alzheimer's disease. Alzheimer's Dementia **8**(1), 1–13 (2012)

20. Iglesias, J.E., et al.: A computational atlas of the hippocampal formation using ex vivo, ultra-high resolution MRI: application to adaptive segmentation of in vivo MRI. Neuroimage **115**, 117–137 (2015)

21. Iglesias, J.E., et al.: A probabilistic atlas of the human thalamic nuclei combining ex vivo MRI and histology. Neuroimage **183**, 314–326 (2018)

22. Marcus, D.S., Wang, T.H., Parker, J., Csernansky, J.G., Morris, J.C., Buckner, R.L.: Open access series of imaging studies (OASIS): cross-sectional MRI data in young, middle aged, nondemented, and demented older adults. J. Cogn. Neurosci. **19**(9), 1498–1507 (2007)

23. Isensee, F., et al.: nnU-Net: a self-configuring method for deep learning-based biomedical image segmentation. Nat. Methods **18**(2), 203–211 (2021)

24. Khandelwal, P., et al.: Automated deep learning segmentation of high-resolution 7 tesla postmortem MRI for quantitative analysis of structure-pathology correlations in neurodegenerative diseases. Imaging Neurosci. **2**, 1–30 (2024). https://direct.mit.edu/imag/article/doi/10.1162/imag_a_00171/120741

25. Zeng, X., et al.: Segmentation of supragranular and infragranular layers in ultra-high resolution 7T ex vivo MRI of the human cerebral cortex. bioRxiv (2023)

26. Jack Jr, C.R., et al.: NIA-AA research framework: toward a biological definition of Alzheimer's disease. Alzheimer's Dementia **14**(4), 535–562 (2018)

27. Jonkman, L.E., et al.: Normal aging brain collection Amsterdam (NABCA): a comprehensive collection of postmortem high-field imaging, neuropathological and morphometric datasets of non-neurological controls. NeuroImage: Clinical (2019)

28. Kotrotsou, A., et al.: Ex vivo MR volumetry of human brain hemispheres. Magn. Reson. Med. **71**(1), 364–374 (2014)
29. Jenkinson, M., et al.: FSL. Neuroimage **62**(2), 782–790 (2012)
30. La Joie, R., et al.: Prospective longitudinal atrophy in Alzheimer's disease correlates with the intensity and topography of baseline tau-PET. Sci. Trans. Med. **12**(524), eaau5732 (2020)
31. Paajanen, T., et al.: CERAD neuropsychological total scores reflect cortical thinning in prodromal Alzheimer's disease. Dement. Geriatr. Cogn. Disord. Extra **3**(1), 446–458 (2013)
32. Ravikumar, S., et al.: Ex vivo MRI atlas of the human medial temporal lobe: characterizing neurodegeneration due to tau pathology. Acta Neuropathologica Commun. **9**(1), 1–14 (2021)
33. Ségonne, F., Pacheco, J., Fischl, B.: Geometrically accurate topology-correction of cortical surfaces using nonseparating loops. IEEE Trans. Med. Imaging **26**(4), 518–529 (2007)
34. Robinson, J.L., et al.: Neurodegenerative disease concomitant proteinopathies are prevalent, age-related and APOE4-associated. Brain **141**(7), 2181–2193 (2018)
35. Sadaghiani, S., et al.: Associations of phosphorylated tau pathology with whole-hemisphere ex vivo morphometry in 7 tesla MRI. Alzheimer's Dementia **19**(6), 2355–2364 (2023)
36. Tisdall, M.D., et al.: Ex vivo MRI and histopathology detect novel iron-rich cortical inflammation in frontotemporal lobar degeneration with tau versus TDP-43 pathology. NeuroImage: Clin. **33**, 102913 (2022)
37. Yushkevich, P.A., et al.: Three-dimensional mapping of neurofibrillary tangle burden in the human medial temporal lobe. Brain **144**(9), 2784–2797 (2021)
38. Billot, B., et al.: SynthSeg: segmentation of brain MRI scans of any contrast and resolution without retraining. Med. Image Anal. **86**, 102789 (2023)
39. Schaefer, A., et al.: Local-global parcellation of the human cerebral cortex from intrinsic functional connectivity MRI. Cereb. Cortex **28**(9), 3095–3114 (2018)
40. Glasser, M.F., et al.: A multi-modal parcellation of human cerebral cortex. Nature **536**(7615), 171–178 (2016)
41. Scholtens, et al.: An MRI von Economo-Koskinas atlas. Neuroimage **170**, 2018 (2018)

Self-supervised Pre-training Tasks for an fMRI Time-Series Transformer in Autism Detection

Yinchi Zhou[1]([✉]), Peiyu Duan[1], Yuexi Du[1], and Nicha C. Dvornek[1,2]

[1] Department of Biomedical Engineering, New Haven, CT, USA
Yinchizhou@gmail.com
[2] Department of Radiology and Biomedical Imaging, Yale University,
New Haven, CT, USA

Abstract. Autism Spectrum Disorder (ASD) is a neurodevelopmental condition that encompasses a wide variety of symptoms and degrees of impairment, which makes the diagnosis and treatment challenging. Functional magnetic resonance imaging (fMRI) has been extensively used to study brain activity in ASD, and machine learning methods have been applied to analyze resting state fMRI (rs-fMRI) data. However, fewer studies have explored the recent transformer-based models on rs-fMRI data. Given the superiority of transformer models in capturing long-range dependencies in sequence data, we have developed a transformer-based self-supervised framework that directly analyzes time-series fMRI data without computing functional connectivity. To address over-fitting in small datasets and enhance the model performance, we propose self-supervised pre-training tasks to reconstruct the randomly masked fMRI time-series data, investigating the effects of various masking strategies. We then fine-tune the model for the ASD classification task and evaluate it using two public datasets and five-fold cross-validation with different amounts of training data. The experiments show that randomly masking entire ROIs gives better model performance than randomly masking time points in the pre-training step, resulting in an average improvement of 10.8% for AUC and 9.3% for subject accuracy compared with the transformer model trained from scratch across different levels of training data availability. Our code is available on GitHub.

Keywords: Autism · fMRI · Transformer · Self-supervised Learning

1 Introduction

Autism Spectrum Disorder (ASD) is a neurodevelopmental condition characterized by intellectual disabilities, impaired social interactions, language impair-

Code available at https://github.com/ycaris/Self-Supervised_fMRI.

Supplementary Information The online version contains supplementary material available at https://doi.org/10.1007/978-3-031-78761-4_14.

ments, and repetitive behaviors. ASD affects a significant portion of the global population, with an estimated prevalence of approximately 1 in 100 children [21]. However, ASD diagnosis is challenging due to the wide range of symptoms and severity, with current diagnostic practices heavily reliant on behavioral and developmental assessments that may be subject to the observer. In addition, the underlying causes of ASD are still unknown. To better characterize ASD phenotypes, functional magnetic resonance imaging (fMRI) has been used to investigate the brain activity of individuals with ASD. fMRI allows for the non-invasive measurement of brain signals through the recording of hemodynamic changes caused by neuronal activity. It provides high spatial resolution and can help locate brain functional activation areas, thereby mapping the connectivity patterns of the brain. Analysis of these signals could enable the identification of biomarkers, early diagnosis, and personalized treatment for ASD.

Prior studies have shown promise in using the whole-brain functional connectivity calculated from the average of the time series of the regions of interest (ROIs) from resting-state fMRI (rs-fMRI) for the characterization and classification of ASD using machine learning [2,7,15,19]. Recently, transformer models [20] have been applied to time-series rs-fMRI to model the dependency among different brain regions across time [1,10]. Transformers are widely used in long sequence language tasks because the attention mechanism is capable of capturing global relationships between distant inputs [4]. In ASD classification applications, Bannadabhavi et al. proposed a hierarchical transformer by learning the relationship of intra- and inter-community among brain regions [1], and Li et al. constructed the positional encoding of the transformer-based model based on the functional connectivity matrix [10]. However, these approaches require the derived functional connectivity matrix as input to the model, rather than leveraging the original time-series data for analysis using the transformer model.

Another challenge of adopting transformer models is that they often require learning on large training data to be successful. The model can first be pre-trained on a large unlabeled dataset, then finetuned on the specific task, improving data efficiency when the labeled data is limited for the target task, as is often the case for fMRI analysis [11,12,14]. The pre-training task usually involves reconstruction of the signal, either in an autoencoder style without masking [11,12] or with random masking of the input [14], akin to masked language modeling [4]. The importance of the design of the pre-training masking strategy for temporal data has been noted in recent work [17]. Considering that the fMRI ROI time series is not only correlated within each individual ROI signal over time, but also between different ROIs as specified by the connectivity of different brain networks, a different masking strategy may be the key to pre-training.

In this work, we investigated the use of different self-supervised pre-training tasks for an fMRI ROI time-series transformer model and evaluated their effects on data efficiency in learning the downstream fine-tuning ASD classification task. We evaluated our method on the public ABIDE [5] and ACE datasets using five-fold cross-validation. The experiments show the effectiveness of transformer

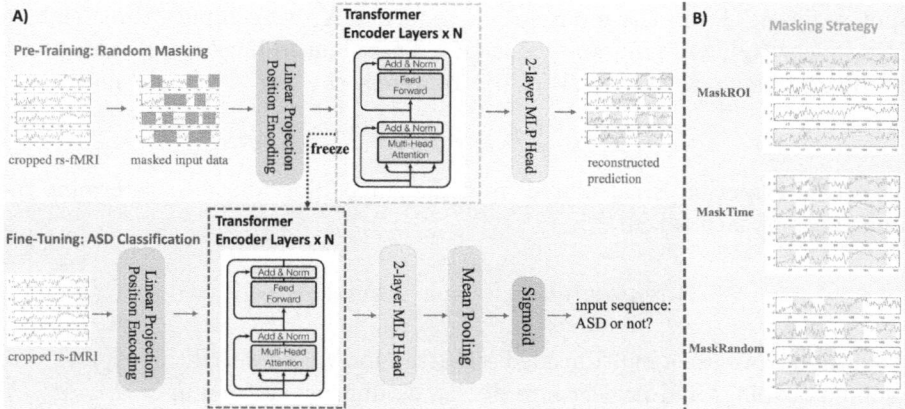

Fig. 1. Framework overview. A) The proposed self-supervised training workflow consists of a pre-training stage and a fine-tuning stage. The cropped rs-fMRI shown is obtained after data augmentation. B) Three masking strategies that are used in the pre-training tasks. Each row is the time-series data for one ROI.

pre-training compared with the scratch transformer model even when trained with 100% of the available data, with the pre-training strategy of randomly masking whole ROIs giving better model performance in the downstream ASD classification task than randomly masking time points in the pre-training step.

2   Methods

2.1   Self-supervised Training Framework Overview

Our self-supervised training framework consists of two stages, the pre-training phase to reconstruct randomly masked time-series fMRI sequences and the fine-tuning phase to train a classifier on top of the transformer encoder for ASD classification (Fig. 1). During the fine-tuning stage, we freeze the parameters of the transformer encoder layers and train an additional two-layer multilayer perceptron (MLP) classifier head to perform the ASD classification.

2.2   Model Architecture

The input to the model is an fMRI time-series of length T from R ROIs. A linear projection layer maps this data to the hidden dimension, and sinusoidal positional encoding creates the embedding for the projected data. The transformer-based model has N transformer encoder layers to extract features from time-series fMRI. The transformer encoder consists of a multi-headed self-attention module, position feed-forward network, residual connectivity, and layer normalization. The self-attention mechanism is the core component of the transformer model, which calculates the importance of individual tokens with respect to the input sequence. From the input tokens sequence X, we compute three embeddings: query (Q), key (K), and value (V). These embeddings are obtained with the formula below where W_Q, W_K, W_V represent the learned weight matrices.

$$Q = XW_Q, \quad K = XW_K, \quad V = XW_V \tag{1}$$

The scaled attention is calculated from the embedding above to determine the importance of each token:

$$Attention(Q, K, V) = softmax(\frac{QK^T}{\sqrt{d_k}})V \tag{2}$$

In our model, we use a multi-headed self-attention where we split Q, K, and V to h sub-embeddings and project into d_k, d_k, d_v dimensions, and $d_k = d_v = d_m/h$, where d_m is the output dimension of the transformer encoder layers:

$$MultiHead(Q, K, V) = Concat(head_1, ..., head_h)W^O \tag{3}$$

$$head_i = Attention(QW_i^Q, KW_i^K, VW_i^V) \tag{4}$$

given the dimension $W_i^Q \in \mathbb{R}^{d_m \times d_k}$, $W_i^K \in \mathbb{R}^{d_m \times d_k}$, $W_i^V \in \mathbb{R}^{d_m \times d_v}$, $W^O \in \mathbb{R}^{hd_v \times d_m}$ Following the transformer encoder layers, a two-layer MLP head is used to reconstruct masked fMRI sequences, and the reconstruction loss is calculated from the cropped sequences and predicted sequences. In the fine-tuning stage, we re-use the same transformer encoder layers to encode the projected inputs, and then train a new MLP head for the downstream task. The output is passed into a sigmoid layer for final classification as follows:

$$Probability = Sigmoid(AvgPool(MLP(ReLU(MLP(Z))))) \tag{5}$$

where Z represents the output from the transformer encoder layer.

2.3   Random Masking in the Pre-training Stage

To investigate the effectiveness of different pre-training tasks, we construct three types of random masks (Fig. 1B). 1) MaskROI: randomly select ROIs and mask the entire time periods of the rs-fMRI sequences. By masking whole ROIs, the model needs to reconstruct the missing indices from non-masked ROIs, enforcing the model to better learn the dependencies among the activity in different brain regions. 2) MaskTime: randomly select time points and mask all ROIs of those time points. This strategy is similar to that used in masked language modeling where random word tokens are masked during pre-training. The model needs to extract information from neighboring time points of specific ROIs. The resulting model may better learn the whole brain signal changes over time. 3) MaskRandom: first randomly select time points and then randomly select ROIs of those time points to mask. This strategy is similar to that used in prior self-supervised pre-training work in fMRI [14]. The masking ratio is randomly set to 0.25 or 0.5. For each strategy, to create the mask, we set the value of the selected indices as zero.

3   Experiments

3.1   Datasets and Pre-Processing

ABIDE I. Autism Brain Imaging Data Exchange (ABIDE) I is a multi-site public dataset including 1112 subjects that were collected from 17 international sites [3,5] (age: 17.0 ± 8.0 years; 948 males and 164 females). It includes rs-fMRI images, T1 structural brain images and phenotypic information for each patient. The preprocesseed data using the Configurable Pipeline for the Analysis of Connectomes (CPAC) was downloaded from [3]. After the quality check, 886 subjects (409 with ASD, 477 healthy controls) were used for model training and testing. The mean time-series for each ROI was extracted using the AAL atlas which parcellated the brain into 116 ROIs [18]. The mean time-series from each ROI was standardized.

ACE. The Autism Centers of Excellence (ACE) public dataset[1] includes comprehensive imaging, behavioral, and other data from a sex-balanced cohort of 526 ASD and neurotypical youth from 4 sites (ages: 13.3 ± 2.9 years). After quality control and filtering for missing data, 282 subjects (140 with ASD, 142 healthy controls) were used in the experiments, including 141 females and 141 males. The data were pre-processed using fMRIPrep, and mean ROI time-series were extracted using the AAL atlas and standardized.

3.2   Experimental Settings

Implementation Details. To increase the effectiveness of pre-training and reduce the overfitting problem on a small dataset, we used a random cropping method as data augmentation to boost the number of training data [6]. Specifically, we randomly cropped 10 sequences with length of $T=64$ time points from the original time-series fMRI data for each subject and re-cropped every epoch to ensure the randomness.

The experiments were performed in PyTorch on an NVIDIA RTX A5000 GPU. In the pre-training stage, the transformer encoder and MLP head was trained for 50000 steps with a batch size of 64, dropout rate of 0.1, weight decay of 10^{-5}, and an initial learning rate of 10^{-4}. We used the cosine learning rate scheduler and the AdamW optimizer. We used $N = 6$ encoder layers with 8 heads for each layer. Mean squared error (MSE) loss between the reconstructed sequences and the cropped sequences was used for optimization. In the fine-tuning stage, the transformer encoder was frozen, and only the MLP classifier was trained using the inner fold data. The model was trained with a batch size of 16, dropout rate of 0.1, weight decay of 10^{-3}, and an initial learning rate of 10^{-4}. Binary cross-entropy loss was optimized for ASD classification. A transformer model for the ASD classification task was also trained from scratch as a baseline. A smaller transformer model with only 2 encoder layers and 4 heads was used

---

[1] Data available from https://nda.nih.gov/edit_collection.html?id=2021.

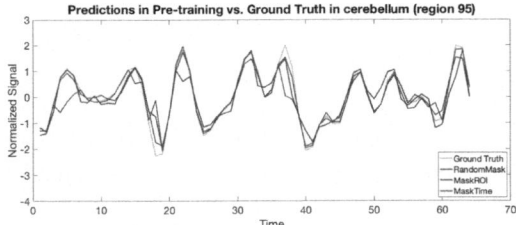

Fig. 2. Visualization of example reconstructed sequences from the left-out testing data using different masking strategies.

Table 1. MSE between reconstructed sequences and ground truth on the left-out testing data across five folds(mean ± std).

Masking Tasks	MSE
MaskRandom	**0.109 ± 0.002**
MaskTime	0.136 ± 0.002
MaskROI	0.136 ± 0.047

when training from scratch to avoid the overfitting problem due to lack of data. To predict the subject label, we applied a sliding inference window on the input sequences of 64 time points from the original time-series fMRI data for each subject and determined the class using the majority voting of input sequences.

Model Evaluation. For model evaluation, we used a nested subject-wise 5-fold cross-validation, stratified by diagnosis labels. In each of the outer 5 folds, 80% of the data was used for training, and 20% of the data was used for testing. Note the outer fold test data was left out in both the pre-training and fine-tuning process to fairly evaluate the model performance on unseen data. The training data used in the pre-training stage was further split into the inner five folds to accommodate the variability of the fine-tuned model. In each inner fold, 80% of the data from the inner fold was used for training and 20% of the data was used for validation to select the best model parameters. The final model evaluation was performed on the left-out test data from five outer folds. In addition, to demonstrate that pre-training improves data efficiency on the downstream task, we used different amounts of training data for ASD classification on the ABIDE dataset in fine-tuning, including 20%, 40%, 60%, 80%, 100% of the training data in each fold. For each of the five testing sets, we obtained a five-fold ensemble result by averaging the model outputs from the inner folds. To evaluate the performance of pre-trained models on out-of-domain data, we first performed pre-training using all of the ABIDE data and then fine-tuned it on the ACE dataset for ASD classification under subject-wise five-fold cross-validation.

Classification model performance was evaluated using receiver operating characteristic (ROC) curve analysis. Significant pairwise differences between models were assessed using paired two-tailed t-tests. For the ABIDE experiments, we used the results from the five cross-validation folds and conducted a two-way repeated measures ANOVA to analyze the effects of the two independent factors of pre-training strategy and the amount of available data during fine-tuning on classification performance. This statistical test aims to find the statistical significance attributable to not only each factor but also the interaction of the two factors. Statistical significance was assessed at the level of p ¡ 0.05 for all tests.

Table 2. Evaluation metrics for ASD classification on ABIDE with different pre-training strategies and different percentages of training data in the fine-tuning stage. The results are summarized from five testing sets. (mean ± std).

Model	AUC				
	20%	40%	60%	80%	100%
Scratch	0.59 ± 0.05	0.61 ± 0.05	0.60 ± 0.05	0.62 ± 0.05	0.65 ± 0.07
MaskRandom	0.61 ± 0.07	0.66 ± 0.04	0.69 ± 0.06	**0.71 ± 0.05**	0.72 ± 0.05
MaskTime	0.57 ± 0.07	0.65 ± 0.05	0.69 ± 0.06	0.70 ± 0.04	0.72 ± 0.04
MaskROI	**0.62 ± 0.07**	**0.67 ± 0.04**	**0.70 ± 0.04**	0.71 ± 0.05	**0.73 ± 0.04**
Model	Subject Accuracy				
	20%	40%	60%	80%	100%
Scratch	0.57 ± 0.06	0.57 ± 0.03	0.59 ± 0.05	0.58 ± 0.03	0.61 ± 0.05
MaskRandom	**0.60 ± 0.04**	**0.62 ± 0.03**	0.64 ± 0.04	**0.66 ± 0.04**	**0.66 ± 0.04**
MaskTime	0.57 ± 0.03	0.60 ± 0.02	**0.65 ± 0.04**	0.65 ± 0.02	0.65 ± 0.02
MaskROI	0.59 ± 0.03	0.61 ± 0.02	0.64 ± 0.04	**0.66 ± 0.06**	**0.66 ± 0.04**

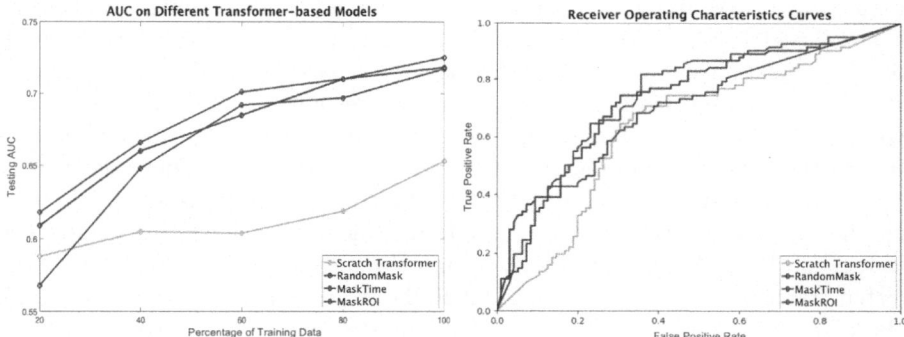

Fig. 3. Left: Testing AUC of ASD classification models learned from different percentages of training data using different pre-training masking strategies. Right: Example ROC curves of one test set using 100% training data.

4 Results

4.1 Pre-training Results with Different Random Masking Strategies

The pre-training task was to reconstruct the masked sequences with $T=64$ time points. The three different masking strategies were shown to be effective visually, where the reconstructed sequences have very similar patterns to the original data (Fig. 2). MSE between the reconstructed sequences and the original data on the left-out testing data for ABIDE across five folds is shown in Table 1. Quantitatively, completely random masking had the lowest average MSE, and MaskROI had higher variability. While MaskRandom resulted in significantly lower MSE

Table 3. Evaluation metrics for ASD classification on ACE dataset (mean ± std). Pre-trained models were learned on all ABIDE data, then fine-tuned on ACE.

Model	AUC	Subject Accuracy
Scratch Transformer	0.56 ± 0.07	0.53 ± 0.04
MaskRandom Finetune	0.61 ± 0.07	0.58 ± 0.05
MaskTime Finetune	0.60 ± 0.05	0.56 ± 0.03
MaskROI Finetune	**0.65 ± 0.04**	**0.61 ± 0.03**

than MaskTime (p ¡ 0.001), no significant difference was found between MaskROI and MaskRandom or MaskTime (p ¿ 0.05).

4.2   Downstream ASD Classification Results

The performance of ASD classification on the ABIDE dataset is shown in Table 2 and Fig. 3. Our results indicate that both AUC and subject accuracy improve as the percentage of training data increases for all models, which is likely to be attributed to the overfitting problem when limited training data is used. We also observe that all pre-training strategies resulted in fine-tuned models with higher average performance than the corresponding scratch model. The two-way repeated measures ANOVA for AUC indicated no significant interaction between the pre-training strategy and the amount of training data (p = 0.152), but showed significant effects for different pre-training tasks (p = 0.002) and the amount of training data (p = 0.023) alone. Specifically, a completely random mask yielded an average AUC improvement of 9.5% compared to the scratch transformer model. Masking all ROIs at randomly selected time points (MaskTime) resulted in an average AUC improvement of 7.5%, while masking all time points at randomly selected ROIs (MaskROI) led to an average AUC improvement of 10.8%. Among different masking strategies, MaskROI consistently achieved the highest AUC and the highest or second-highest accuracy across all percentages. This may be attributed to the model's enhanced ability to learn the relationships between different brain regions, corresponding to better modeling of brain connectivity, which has shown to be widely affected in ASD patients [8,9,13,16]. Subject accuracy also improved with the use of the pre-trained model across all percentages. Differences in masking strategies were evident in sensitivity and specificity metrics (Supplementary Table S1). MaskROI exhibited the highest sensitivity yet lowest specificity, but also it was the only model that produced overall average sensitivity and specificity that were both greater than 0.5.

The performance of ASD classification on the ACE dataset is shown in Table 3. Results showed that the fine-tuned models using all masking strategies outperformed the scratch transformer model for ASD classification, indicating the effectiveness of pre-training even with different datasets. The small size of the ACE dataset causes overfitting of the transformer model trained from scratch to the training data, leading to low AUC and subject accuracy on the test set. By using a pretrained encoder from the ABIDE dataset, we can train a classification head to mitigate the overfitting problem. Notably, the fine-tuned model with MaskROI achieved the highest AUC and accuracy and was the only pre-training method that resulted in a significantly higher performance than the scratch model ($p = 0.006$ for AUC, $p = 0.040$ for accuracy). Furthermore, the MaskROI model performed significantly better than the MaskTime model ($p = 0.048$ for AUC, $p = 0.020$ for accuracy). These results highlight the importance of the choice of the pre-training task.

5 Conclusion

In this work, we proposed a transformer-based self-supervised pre-training and fine-tuning framework for ASD classification using time-series fMRI data. The framework includes a pre-training stage where input sequences are masked using three different strategies: randomly masking time points, randomly masking ROIs, and completely random masking. We trained and evaluated our method on two datasets and observed significant improvements compared to the scratch transformer model without pre-training. Our results indicate that the performance differs among the three masking strategies used during pre-training. Randomly masking time points underperforms compared to the other two strategies while masking entire ROIs achieves the best performance on both the ABIDE and ACE datasets. Future work will focus on incorporating multimodal information into the current framework to further enhance classification accuracy.

Acknowledgements. This research is supported in part by the National Institute of Neurological Disorders and Stroke (NINDS) of the National Institutes of Health grant R01NS035193.

Disclosure of Interests. The authors have no competing interests to declare that are relevant to the content of this article.

References

1. Bannadabhavi, A., Lee, S., Deng, W., Ying, R., Li, X.: Community-aware transformer for autism prediction in fMRI connectome. In: Greenspan, H., et al. (eds.) Medical Image Computing and Computer Assisted Intervention – MICCAI 2023: 26th International Conference, Vancouver, BC, Canada, October 8–12, 2023, Proceedings, Part VIII, pp. 287–297. Springer Nature Switzerland, Cham (2023). https://doi.org/10.1007/978-3-031-43993-3_28
2. Chen, C.P., et al.: Diagnostic classification of intrinsic functional connectivity highlights somatosensory, default mode, and visual regions in autism. NeuroImage: Clin. **8**, 238–245 (2015)
3. Craddock, C., et al.: The neuro bureau preprocessing initiative: open sharing of preprocessed neuroimaging data and derivatives. Front. Neuroinform. **7**(27), 5 (2013)
4. Devlin, J., Chang, M.W., Lee, K., Toutanova, K.: BERT: pre-training of deep bidirectional transformers for language understanding. arXiv preprint arXiv:1810.04805 (2018)

5. Di Martino, A., et al.: The autism brain imaging data exchange: towards a large-scale evaluation of the intrinsic brain architecture in autism. Mol. Psych. **19**(6), 659–667 (2014)

6. Dvornek, N.C., Ventola, P., Pelphrey, K.A., Duncan, J.S.: Identifying autism from resting-state fMRI using long short-term memory networks. In: Wang, Q., Shi, Y., Suk, H.-I., Suzuki, K. (eds.) MLMI 2017. LNCS, vol. 10541, pp. 362–370. Springer, Cham (2017). https://doi.org/10.1007/978-3-319-67389-9_42

7. Heinsfeld, A.S., Franco, A.R., Craddock, R.C., Buchweitz, A., Meneguzzi, F.: Identification of autism spectrum disorder using deep learning and the abide dataset. NeuroImage: Clin. **17**, 16–23 (2018)

8. Hull, J.V., Dokovna, L.B., Jacokes, Z.J., Torgerson, C.M., Irimia, A., Van Horn, J.D.: Resting-state functional connectivity in autism spectrum disorders: a review. Front. Psych. **7**, 205 (2017)

9. Kana, R.K., Keller, T.A., Minshew, N.J., Just, M.A.: Inhibitory control in high-functioning autism: decreased activation and underconnectivity in inhibition networks. Biol. Psych. **62**(3), 198–206 (2007)

10. Li, W., Wang, S., Liu, G.: Transformer-based model for fMRI data: abide results. In: 2022 7th International Conference on Computer and Communication Systems (ICCCS), pp. 162–167. IEEE (2022)

11. Malkiel, I., Rosenman, G., Wolf, L., Hendler, T.: Pre-training and fine-tuning transformers for fMRI prediction tasks. arXiv preprint arXiv:2112.05761 **105** (2021)

12. Malkiel, I., Rosenman, G., Wolf, L., Hendler, T.: Self-supervised transformers for fMRI representation. In: International Conference on Medical Imaging with Deep Learning, pp. 895–913. PMLR (2022)

13. Müller, R.A., Shih, P., Keehn, B., Deyoe, J.R., Leyden, K.M., Shukla, D.K.: Underconnected, but how? A survey of functional connectivity MRI studies in autism spectrum disorders. Cereb. Cortex **21**(10), 2233–2243 (2011)

14. Ortega Caro, J., et al.: BrainLM: a foundation model for brain activity recordings. bioRxiv (2023)

15. Plitt, M., Barnes, K.A., Martin, A.: Functional connectivity classification of autism identifies highly predictive brain features but falls short of biomarker standards. NeuroImage: Clin. **7**, 359–366 (2015)

16. Rane, P., Cochran, D., Hodge, S.M., Haselgrove, C., Kennedy, D.N., Frazier, J.A.: Connectivity in autism: a review of MRI connectivity studies. Harv. Rev. Psych. **23**(4), 223–244 (2015)

17. Tong, Z., Song, Y., Wang, J., Wang, L.: VideoMAE: masked autoencoders are data-efficient learners for self-supervised video pre-training. Adv. Neural. Inf. Process. Syst. **35**, 10078–10093 (2022)

18. Tzourio-Mazoyer, N., et al.: Automated anatomical labeling of activations in SPM using a macroscopic anatomical parcellation of the MNI MRI single-subject brain. Neuroimage **15**(1), 273–289 (2002)

19. Van Dijk, K.R., Hedden, T., Venkataraman, A., Evans, K.C., Lazar, S.W., Buckner, R.L.: Intrinsic functional connectivity as a tool for human connectomics: theory, properties, and optimization. J. Neurophysiol. **103**(1), 297–321 (2010)

20. Vaswani, A., et al.: Attention is all you need. In: Advances in Neural Information Processing Systems, vol. 30 (2017)

21. Zeidan, J., et al.: Global prevalence of autism: a systematic review update. Autism Res. **15**(5), 778–790 (2022)

Is Your Style Transfer Doing Anything Useful? An Investigation into Hippocampus Segmentation and the Role of Preprocessing

Hoda Kalabizadeh[1]([⊠]), Ludovica Griffanti[2], Pak-Hei Yeung[3], Natalie Voets[4], Grace Gillis[2], Clare Mackay[2], Ana IL Namburete[1], Nicola K. Dinsdale[1], and Konstantinos Kamnitsas[5]

[1] Department of Computer Science, University of Oxford, Oxford, UK
hoda.kalabizadeh@cs.ox.ac.uk
[2] Department of Psychiatry, University of Oxford, Oxford, UK
[3] College of Computing and Data Science, Nanyang Technological University, Singapore, Singapore
[4] Nuffield Department of Clinical Neurosciences, University of Oxford, Oxford, UK
[5] Department of Engineering Science, University of Oxford, Oxford, UK

Abstract. Brain atrophy assessment in MRI, particularly of the hippocampus, is commonly used to support diagnosis and monitoring of dementia. Consequently, there is a demand for accurate automated hippocampus quantification. Most existing segmentation methods have been developed and validated on research datasets and, therefore, may not be appropriate for clinical MR images and populations, leading to potential gaps between dementia research and clinical practice. In this study, we investigated the performance of segmentation models trained on research data that were *style-transferred* to resemble clinical scans. Our results highlighted the importance of intensity normalisation methods in MRI segmentation, and their relation to domain shift and style-transfer. We found that whilst normalising intensity based on min and max values, commonly used in generative MR harmonisation methods, may *create* a need for style transfer, Z-score normalisation effectively maintains style consistency, and optimises performance. Moreover, we show for our datasets spatial augmentations are more beneficial than style harmonisation. Thus, emphasising robust normalisation techniques and spatial augmentation significantly improves MRI hippocampus segmentation.

Keywords: Style Transfer · Hippocampus Segmentation · Dementia

---

N. K. Dinsdale and K. Kamnitsas—Equal contribution.

Supplementary Information The online version contains supplementary material available at https://doi.org/10.1007/978-3-031-78761-4_15.

1   Introduction

Many neurodegenerative diseases cause volumetric atrophy in the region of the hippocampus [1], including Alzheimer's disease (AD). AD is clinically characterised by a progressive decline in cognitive function with diagnosis and monitoring of the disease commonly including the assessment of hippocampal atrophy in brain MRI scans [2]. Specifically, the volume of the hippocampus, is often measured, either through manual or automated segmentation.

Manual segmentation requires large amounts of time and expert knowledge and suffers from inter-rater variability. Therefore, there is a demand for accurate automated hippocampus segmentation methods [3]. Among different types of techniques, deep learning (DL) based methods show great promise for the segmentation of the hippocampus, outperforming traditional atlas-based approaches [4,5]. However, training CNNs generally requires the availability of manual labels, limiting the applicability in clinical practice. Furthermore, due to differences in image acquisitions and patient demographics, models trained on research datasets are unlikely to generalise to clinical populations. Therefore, there is a need to overcome this *domain shift* between the source (research) and target (clinical) dataset, enabling the development of segmentation models for clinical scans without requiring segmentation labels.

To address general domain shift, data augmentation is a commonly used technique for artificially enhancing the diversity of training data, to increase model generalisability and robustness. Augmentation has been shown to improve downstream segmentation performance across brain imaging studies [6]. However, augmentation requires the identification and modelling of differences between domains, which is non-trivial in the presence of varying populations and scanner technologies. Another related field of research is image-to-image (I2I) translation, which is an approach that aims to learn the mapping between different visual domains, mostly based on generative models. For instance, Pix2pix [7] utilities a conditional GAN to map between image domains, but relies on pixel-to-pixel correspondence, limiting its applicability to MR images from different sites. CycleGAN [8] overcomes the need for paired data using cycle consistency.

MR harmonisation approaches, e.g., [9] are based on I2I methods, aiming to overcome the style-based domain shifts associated with differing acquisition scanners while maintaining the underlying anatomy.

Therefore, in this study, we aim to investigate which techniques are effective for overcoming the *domain shift* between our source (research) dataset and target (clinical) dataset for the task of MR hippocampus segmentation. Our contributions are as follow:

- We demonstrate the use of a 2-stage pipeline for generating style-transferred images that are subsequently used to train a hippocampus segmentation model.
- We explore the impact on downstream hippocampus segmentation performance of different preprocessing and augmentation approaches.
- We show that the use of appropriate normalisation (*i.e.* Z-score normalisation) and spatial augmentation (*i.e.* paired affine registration) can lead to

substantial improvements on downstream hippocampus segmentation performance, even without a sophisticated style transfer pipeline.

The findings of this study may shed light on the importance of developing a robust preprocessing pipeline for MR hippocampus segmentation in future studies.

2 Methods

To overcome the domain shift between the research and clinical data, we implemented a 2-stage approach, formed of a style transfer (ST) network followed by a segmentation network. A schematic of this pipeline is shown in Fig. 1. We assumed access to a source (research) dataset, $\mathcal{D}_s = \{\boldsymbol{X}_s, \boldsymbol{Y}_s\}$, and an unlabelled target (clinical) dataset, $\mathcal{D}_t = \{\boldsymbol{X}_t\}$, to first train a style transfer model that generates source images in the style of target images, $\tilde{\boldsymbol{X}}_s = G(\boldsymbol{X}_s, \boldsymbol{X}_t)$, following which we used the ST images to train a segmentation model $f(\tilde{\boldsymbol{X}}_s, \boldsymbol{Y}_s)$, such that the performance for $\mathcal{D}_t$ is maximised.

2.1 Style Transfer: Style-Encoding GAN

We utilised the Style-Encoding GAN (SE-GAN) [9] for our style transfer network. Similarly to StarGANv2 [10], it is formed of a single generator (G), discriminator (D), mapping network (M) and a style encoder (E). During training, SE-GAN trains G to generate diverse images corresponding to a single image slice $\boldsymbol{x} \in \boldsymbol{X}$ using a style code c, provided by either M or E. Consequently, the generator G translates an input image, $\boldsymbol{x}$, into an output image, $\tilde{\boldsymbol{x}} = G(\boldsymbol{x}, c)$, that is reflective of the style of c. To validate the successful injection of c into the output image $\tilde{\boldsymbol{x}}$, E is used to extract the style code from images. The style code is a 1×64 vector, allowing E to produce diverse style codes from different images. Moreover, the discriminator D learns to classify images as real or fake, as produced by $G(\boldsymbol{x}, c)$. In our experiments, G is used to synthesise output images $\tilde{\boldsymbol{x}}_s$ based on source images $\boldsymbol{x}_s$ that are reflecting the style c extracted from various reference images in $\boldsymbol{X}_t$. Finally, 3D volumes can be reconstructed by stacking the 2D slices. The network is trained using the loss introduced in [10], formed of an adversarial loss L_{GAN}, cycle consistency loss L_{cyc}, style reconstruction loss L_{sty} and diversification loss L_{div}, weighted by λ_{cyc}, λ_{sty} and λ_{div} respectively, resulting in the following objective function:

$$L(G, M, E, D) = L_{GAN} + \lambda_{cyc}L_{cyc} + \lambda_{sty}L_{sty} - \lambda_{div}L_{div} \tag{1}$$

2.2 Segmentation: U-Net

The second stage of the framework is the training of the segmentation network, f, for which we used a 3D U-Net [11]. The network is trained with a Dice loss, L_{dice}, using the style transformed source data such that:

$$L_{seg}(\tilde{\boldsymbol{X}}_s, \boldsymbol{Y}_s) = L_{dice}(f(\tilde{\boldsymbol{X}}_s), \boldsymbol{Y}_s). \tag{2}$$

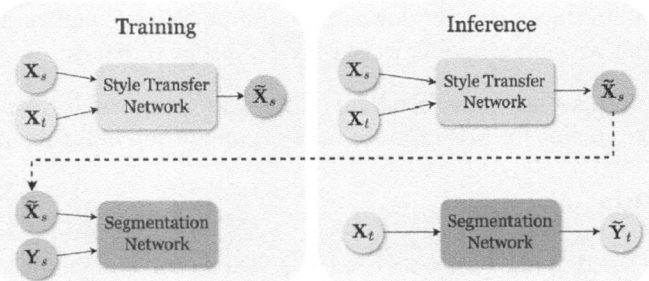

Fig. 1. A schematic of the proposed 2-stage pipeline.

2.3　Investigating Preprocessing

Registration: To mitigate content shift, defined as variations in anatomical alignment between brain scans, we investigated the impact of registration on both style transfer and downstream segmentation performance. To this end, we conducted registration, using both 6 (rigid-body) and 12 (affine) degrees of freedom (DOF) and compared two main approaches. (1) **MNI-Reg**: A spatial harmonisation approach, where both source and target images are registered to a standard space, (2) **Paired-Reg**: A spatial augmentation approach, where every source image is registered to every target image.

Normalisation: In most ST models intensity normalisation is performed during training, through linear scaling of the range of intensities from [Min, Max] to a pre-defined range such as [0,1], which we call Min-Max normalisation. This approach, however, is often unsuitable for MRI images as they can have varying intensity distributions from different acquisitions, leading to inconsistent normalisation results. Additionally, intensity outliers common in MRI data can skew normalisation. We investigated the performance of Min-Max normalisation and explored the impact of applying Z-score normalisation on a per-subject basis.

3　Experimental Setup

3.1　Datasets

Research Dataset: The HarP dataset [12] was used as the labelled *research dataset*, consisting of 135 T1-weighted MRI volumes (cognitively normal controls, MCI and AD patients) from a range of scanners, and corresponding hippocampus masks [13]. All MRIs were registered to MNI-space.

Clinical Dataset: We used a dataset from the Oxford Brain Health Clinic (OBHC) [14] as our *clinical dataset*, representing the unlabelled target domain. It includes 29 patients referred to a memory clinic, who agreed to the use of data for research. The lack of strict inclusion criteria typical of a dementia research study, makes this dataset representative of real-world memory clinic patients. The scans were collected using a 3T Siemens scanner. Hippocampi were manually annotated by a clinician. BHC labels were used only for model evaluation, not for training.

Figure 2 compares the hippocampal volumes between the research (HarP) and clinical (BHC) populations. It can be seen that generally the research population have larger hippocampal volumes than the clinical group. This difference can probably be attributed to dementia research typically recruiting patients that tend to be younger and have less hippocampal atrophy [15].

3.2   Preprocessing

For anonymisation, the BHC scans were brain extracted and thus we performed brain extraction on the HarP dataset. N4 bias field correction was used to correct for low-frequency intensity non-uniformity. Images were split into left and right hemispheres for training. Data registration followed Sect. 2.3.

3.3   Implementation Details

For training the ST network, 132 HarP images were used as the source images, and 20 randomly selected BHC images were used as the references, using a learning rate of 10^{-4} and the Adam optimiser. For Eq. 1, we set $\lambda_{cyc} = 10$, $\lambda_{sty} = 1$ and $\lambda_{div} = 1$, as suggested by [9]. Moreover, 100 HarP images were used for training and then the trained ST network was used to generate a style-transferred image for each HarP image (N=32) in the style of each BHC image (N=20) resulting in 640 style-transferred images, which were used to train the segmentation model.

The chosen U-Net architecture network had four downsampling and upsampling layers, whereby each layer was formed of a convolutional layer, a ReLU activation function and a batch normalisation layer. The depth, defined as the number of convolutions, doubled between each layer, starting with 4. The U-Nets were trained using a learning rate of 10^{-3}, and the Adam optimiser. The training was conducted using 3-fold cross-validation and tested on 9 BHC patients

Table 1. DSC for segmentation methods on HarP (Source) and OBHC (Target). N is the number of test hippocampi (i.e., 2× number of patients). * UDA test sizes were smaller due to training on a sample of unlabelled OBHC.

Method	HarP (N=64)	OBHC (N=58)
FreeSurfer	0.701 ± 0.049	0.625 ± 0.217
SynthSeg	0.801 ± 0.045	0.732 ± 0.070
FIRST	0.810 ± 0.031	0.758 ± 0.116
Hippodeep	0.829 ± 0.031	0.752 ± 0.062
U-Net	0.854 ± 0.048	0.670 ± 0.171
Basic Aug	0.860 ± 0.041	**0.783 ± 0.051**
MRI Aug	0.854 ± 0.040	0.764 ± 0.055
UDA*	**0.863 ± 0.041**	0.742 ± 0.078

(18 hippocampi) that were not used during the training or validation. Training was conducted using an Nvidia A10 GPU. The ST and segmentation networks required an average training time of 35 h and 12 h, respectively. However, once training was complete, the segmentation model's testing, or inference, took only a few seconds per scan, making it suitable for clinical applications.

4   Results and Discussion

4.1   Domain Shift

First, to establish the domain shift between the datasets, we tested publicly available out-of-the-box (OOB) tools: FSL FIRST [16], FreeSurfer [17], Synth-Seg [5], Hippodeep [18], as well as U-Net based approaches: a U-Net trained solely on HarP, basic augmentation (affine transforms, flips, noise, intensity changes), MRI-specific augmentation (motion, bias field). We also compared with adversarial unsupervised domain adaptation, approach shown potent for tackling domain shift in medical imaging [19], and specifically the model developed in [20] (UDA).

Table 1 shows the dice score (DSC) values for the different approaches. The OOB models all performed better on our research population compared to our clinical population, achieving maximum DSC of 0.829 and 0.758, respectively. In particular, FreeSurfer and FIRST had instances of complete failure for the OBHC data (DSC = 0). For most patients, UDA was comparable to the augmentation methods, however, when examining the worst-case scenarios, UDA led to particularly low dice scores for certain individuals. Data augmentation, thus, proved to be the most effective approach for performance enhancement. These findings demonstrate the limitations of existing methods and highlight the potential value of exploring more sophisticated data augmentation approaches.

4.2   Registration

We then explored the effect of the choice of registration approach. Figure 2 shows the effect of registration on the hippocampal volumes: as rigid registration only

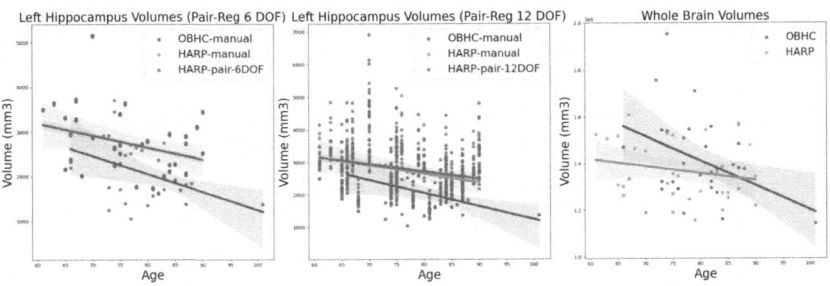

Fig. 2. Volume against age plots for HarP and OBHC: left hippocampus volumes for Paired-Reg 6 DOF (left), 12 DOF (middle), and whole brain volumes (right).

involves translation and rotation for brain alignment, there is no change between the original HarP volumes and those registered to OBHC. However, affine registration performs global scaling, resulting in a slight increase in volume of the registered HarP hippocampi. Although the OBHC registration targets have distinctly smaller hippocampal volumes, they have, on average, larger whole-brain volumes, leading to larger hippocampi in the registered HARP images.

4.3   Style Transfer Vs Normalisation

Table 2 shows the segmentation performance for models trained on Z-score and Min-Max normalised data, and tested on the 9 unseen OBHC patients (18 hippocampi), for a range of evaluation metrics, namely the Dice score (DSC), Hausdorff distance (HD), and Relative Absolute Volume Difference (RAV). As a supervised benchmark, we trained directly on the OBHC dataset (20 labelled patients), which achieved an average DSC of 0.744. By comparison, simply training on the Min-Max normalised HarP dataset achieved a mean DSC of 0.616, clearly indicating presence of a domain shift.

The results of training with the U-Net on the HarP dataset with the different registration schemes, normalisation schemes and the use of ST can then be seen. MNI-Reg-6 ST led to a 4% increase in performance, with a DSC of 0.651. Z-score normalisation outperformed Min-Max normalisation across the experiments. Without Z-score normalisation, a noticeable style shift exists, which can be slightly mitigated by training on style-transferred images (MNI-Reg-6 ST). However, implementing Z-score normalisation effectively reduces this style shift, increasing performance to levels similar to a model trained on target data, while the benefits offered by style transfer are reduced. Following this, the impact of mitigating content shift using affine registration was evaluated, specifically employing the paired registration approach (Table 2). A significant increase in segmentation performance is observed through augmenting the data with paired registration (Paired-Reg-12), achieving an average DSC of 0.780 without ST and 0.787 with. Standard augmentations further improved the performance, achiev-

Table 2. Segmentation results using different normalisation and registrations. N is the number of train hippocampi (i.e. 2× number of patients)

Train Data	Train N	Norm	Min DSC↑	Avg DSC↑	95 % HD↓	RAV↓
OBHC	40	Z-score	0.617	0.744±0.016	3.865±1.357	18.388±1.694
HarP	64	Min-Max	0.480	0.616±0.037	5.264±0.970	97.972±9.456
MNI-Reg-6 ST	1,280	Min-Max	0.521	0.651±0.015	4.036±0.06	83.033±2.037
HarP	64	Z-score	0.674	0.746±0.009	6.013±1.297	26.326±1.856
MNI-Reg-6 ST	1,280	Z-score	0.688	0.757±0.008	5.072±1.391	20.736±1.906
Paired-Reg-12	1,280	Z-score	0.703	0.780±0.005	3.452±0.237	20.009±0.913
Paired-Reg-12 ST	1,280	Z-score	0.672	0.787±0.004	3.169±0.250	23.794±4.033
Paired-Reg-12 ST + Aug	1,280	Z-score	0.719	0.797±0.003	3.133±0.143	27.509±3.724

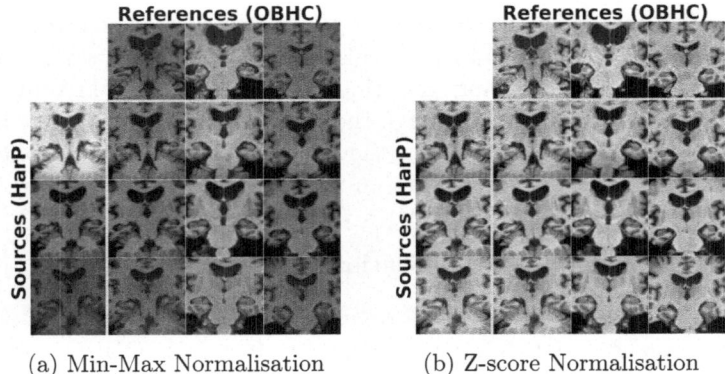

(a) Min-Max Normalisation   (b) Z-score Normalisation

Fig. 3. Generated ST images for a given source (first column) and reference (first row), using a) Min-Max and b) Z-score normalisation.

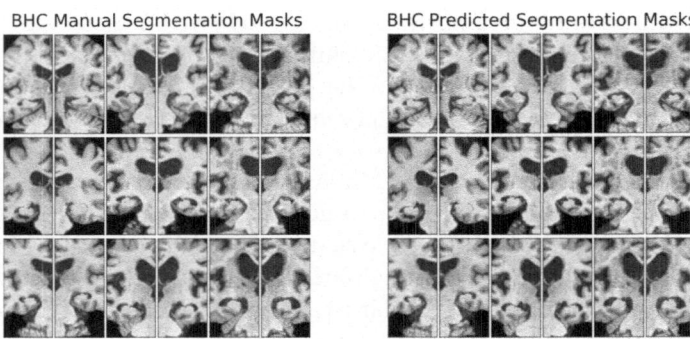

Fig. 4. Manual and predicted segmentation masks for the OBHC test data.

ing the highest DSC of 0.797 (Paired-Reg-12 ST + Aug). The source, reference and style-transferred images, generated by ST networks trained on affine paired registered images (Paired-Reg-12 ST) using both normalisation approaches, have been visualised in Fig. 3. The figures reveal that Min-Max normalisation tends to highlight the style transfer effect more visibly than Z-score normalisation. This difference arises because Min-Max normalisation is sensitive to extreme values in MRI data, which can distort the results. In contrast, Z-score normalisation is more robust to such outliers. Thus, the differences between the figures likely reflect variations in intensity ranges rather than style transfer performance. This difference is further demonstrated by the intensity distributions plotted in the Supplementary Material. Figure 4 provides a qualitative comparison between the manual segmentations and the best performing model predictions (Paired-Reg-12 ST with augmentations) on the OBHC test data.

5   Conclusion

In conclusion, we implemented a 2-stage pipeline consisting of a ST and a segmentation network. Our experimental findings underscored the significance of normalisation methods in MRI augmentation and segmentation tasks. While experiments with Min-Max normalisation may suggest a style shift and the potential benefits of style transfer, this interpretation may be misleading and is a result of inappropriate normalisation. Our findings indicate that Z-score normalisation negates the necessity for style transfer by effectively maintaining style consistency in MRI data, thereby optimising segmentation performance directly. Moreover, for the task of hippocampus segmentation, our results demonstrate that mitigating the content shift using a spatial augmentation approach (i.e. Paired-Reg 12 DOF) can be far more beneficial than a spatial harmonisation approach, such as aligning all images to MNI. The improved performance may be attributed to the spatial diversity introduced by the augmentation, which enhances segmentation robustness. Thus, prioritising robust normalisation techniques and appropriate spatial augmentation can lead to substantial improvements in the generalisability of MRI segmentation. Future studies may, thus, benefit from considering spatial augmentation, akin to those currently employed in style transfer, to achieve further improvements in hippocampus segmentation performance.

Acknowledgments. The authors are grateful for support from: the University of Oxford Department of Computer Science Scholarship (HK), the Bill and Melinda Gates Foundation (NKD, AILN) and the Presidential Postdoctoral Fellowship (Nanyang Technological University) (PHY). We are grateful to the operations team of the OBHC. The OHBC data collection and analysis is supported by the NIHR Oxford Health Biomedical Research Centre (NIHR203316) - a partnership between the University of Oxford and Oxford Health NHS Foundation Trust, the NIHR Oxford Cognitive Health Clinical Research Facility, and the Wellcome Centre for Integrative Neuroimaging (203139/Z/16/Z, 203139/A/16/Z). The views expressed are those of the author(s) and not necessarily those of the NIHR or the Department of Health and Social Care. For the purpose of open access, the authors have applied a CC BY public copyright licence to any Author Accepted Manuscript version arising from this submission.

Disclosure of Interests. The authors have no competing interests to declare.

References

1. Minkova, L., Habich, A., Peter, J., Kaller, C.P., Eickhoff, S.B., Klöppel, S.: Gray matter asymmetries in aging and neurodegeneration: a review and meta-analysis. Hum. Brain Mapping **38**, (12), 5890–5904 (2017). ISSN: 1097–0193. https://doi.org/10.1002/hbm.23772
2. McKhann, G.M., Knopman, D.S., Chertkow, H., et al.: The diagnosis of dementia due to Alzheimer's disease: recommendations from the national institute on aging-Alzheimer's association workgroups on diagnostic guidelines for Alzheimer's disease. Alzheimer's Dementia: J. Alzheimer's Assoc. **7**(3), 263–269 (2011). ISSN:1552–5260. https://doi.org/10.1016/j.jalz.2011.03.005

3. Balboni, E., Nocetti, L., Carbone, C., et al.: The impact of transfer learning on 3D deep learning convolutional neural network segmentation of the hippocampus in mild cognitive impairment and Alzheimer disease subjects. Hum. Brain Mapping **43**(11), 3427–3438 (2022). ISSN: 1097–0193. https://doi.org/10.1002/hbm.25858

4. Dinsdale, N.K., Jenkinson, M., Namburete, A.I.L.: Spatial warping network for 3D segmentation of the hippocampus in MR images. In: Shen, D., et al. (eds.) Medical Image Computing and Computer Assisted Intervention – MICCAI 2019: 22nd International Conference, Shenzhen, China, October 13–17, 2019, Proceedings, Part III, pp. 284–291. Springer International Publishing, Cham (2019). https://doi.org/10.1007/978-3-030-32248-9_32

5. Billot, B., et al.: SynthSeg: Segmentation of brain MRI scans of any contrast and resolution without retraining. Med. Image Anal. **86**, 102789 (2023). https://doi.org/10.1016/j.media.2023.102789

6. Garcea, F., Serra, A., Lamberti, F., Morra, L.: Data augmentation for medical imaging: a systematic literature review. Comput. Biol. Med. **152**, 106391 (2023). https://doi.org/10.1016/j.compbiomed.2022.106391

7. Isola, P., Zhu, J.-Y., Zhou, T., Efros, A.A.: Image-to-image translation with conditional adversarial networks, version: 1 (2016). https://doi.org/10.48550/arXiv.1611.07004

8. Yang, H., et al.: Unpaired brain MR-to-CT synthesis using a structure-constrained CycleGAN. In: Stoyanov, D., et al. (eds.) Deep Learning in Medical Image Analysis and Multimodal Learning for Clinical Decision Support: 4th International Workshop, DLMIA 2018, and 8th International Workshop, ML-CDS 2018, Held in Conjunction with MICCAI 2018, Granada, Spain, September 20, 2018, Proceedings, pp. 174–182. Springer International Publishing, Cham (2018). https://doi.org/10.1007/978-3-030-00889-5_20

9. Liu, M., Zhu, A.H., Maiti, P., et al.: Style transfer generative adversarial networks to harmonize multisite MRI to a single reference image to avoid overcorrection. Hum. Brain Mapping **44**(14), 4875–4892 (2023). ISSN: 1097–0193. https://doi.org/10.1002/hbm.26422

10. Choi, Y., Choi, M., Kim, M., Ha, J.-W., Kim, S., Choo, J.: StarGAN: unified generative adversarial networks for multi-domain image-to-image translation. In: 2018 IEEE/CVF Conference on Computer Vision and Pattern Recognition, pp. 8789–8797 (2018). ISSN: 2575–7075. https://doi.org/10.1109/CVPR.2018.00916

11. Ronneberger, O., Fischer, P., Brox, T.: U-Net: convolutional networks for biomedical image segmentation (2015). https://doi.org/10.48550/arXiv.1505.04597

12. Boccardi, M., Bocchetta, M., Apostolova, L.G., et al.: Delphi definition of the EADC-ADNI harmonized protocol for hippocampal segmentation on magnetic resonance. Alzheimer's Dementia **11**(2), 126–138 (2015). ISSN: 1552–5279. https://doi.org/10.1016/j.jalz.2014.02.009

13. Boccardi, M., Bocchetta, M., Morency, F.C., et al.: Training labels for hippocampal segmentation based on the EADC-ADNI harmonized hippocampal protocol. Alzheimer's & Dementia **11**(2), 175–183 (2015). ISSN: 1552–5279. https://doi.org/10.1016/j.jalz.2014.12.002

14. O'Donoghue, M.C., Blane, J., Gillis, G., et al.: Oxford brain health clinic: protocol and research database. BMJ Open **13**(8), e067808 (2023). Publisher: British Medical Journal Publishing Group Section: Neurology, ISSN: 2044–6055. https://doi.org/10.1136/bmjopen-2022-067808

15. Thorogood, A., et al.: Consent recommendations for research and international data sharing involving persons with dementia. Alzheimer's Dementia **14**(10), 1334–1343 (2018). https://doi.org/10.1016/j.jalz.2018.05.011

16. Patenaude, B., Smith, S.M., Kennedy, D.N., Jenkinson, M.: A Bayesian model of shape and appearance for subcortical brain segmentation. Neuroimage **56**(3), 907–922 (2011). https://doi.org/10.1016/j.neuroimage.2011.02.046

17. Fischl, B., et al.: Whole Brain Segmentation. Neuron **33**(3), 341–355 (2002). https://doi.org/10.1016/S0896-6273(02)00569-X

18. Thyreau, B., Sato, K., Fukuda, H., Taki, Y.: Segmentation of the hippocampus by transferring algorithmic knowledge for large cohort processing. Med. Image Anal. **43**, 214–228 (2018). https://doi.org/10.1016/j.media.2017.11.004

19. Kamnitsas, K., Baumgartner, C., Ledig, C., et al.: Unsupervised domain adaptation in brain lesion segmentation with adversarial networks. In: Niethammer, M., Styner, M., Aylward, S., et al., Eds., Information Processing in Medical Imaging, Series: Lecture Notes in Computer Science, pp. 597–609. Springer International Publishing, Cham (2017). ISBN: 978-3-319-59050-9. https://doi.org/10.1007/978-3-319-59050-9_47

20. Dinsdale, N.K., Jenkinson, M., Namburete, A.I.L.: Deep learning based unlearning of dataset bias for MRI harmonisation and confound removal. NeuroImage **228**, 117–689 (2021). ISSN: 1053–8119. https://doi.org/10.1016/j.neuroimage.2020.117689

GAMing the Brain: Investigating the Cross-Modal Relationships Between Functional Connectivity and Structural Features Using Generalized Additive Models

Arunkumar Kannan[1]([envelope]), Brian Caffo[2], and Archana Venkataraman[3]

[1] Department of Electrical and Computer Engineering, Johns Hopkins University, Baltimore, USA
akannan7@jhu.edu
[2] Department of Biostatistics, Johns Hopkins University, Baltimore, USA
bcaffo1@jhu.edu
[3] Department of Electrical and Computer Engineering, Boston University, Boston, USA
archanav@bu.edu

Abstract. Functional connectivity, reflecting synchronized brain activity across distinct regions, is crucial for understanding cognitive processes. Despite the recent interest in exploring the relationship between functional connectivity and structural brain features, understanding the precise link remains challenging. We propose a novel analysis method that integrates structural factors-such as anatomical morphology summaries, voxel intensity, diffusion-weighted information, and geographic distance to explain variation in functional connectivity. Our method employs generalized additive model (GAM), leveraging region-pair or vertex-pair information, while accommodating individual subject differences in both template and subject spaces. Furthermore, we assess repeatability via the so called discriminability of subjects under our approach, quantifying the probability of similarities between measurements for the same subject versus different subjects. Utilizing data from the Human Connectome Project, we analyze brain connectivity in twin pairs and non-twin pairs to evaluate the repeatability of model-based connectivity patterns estimated via GAMs. Our findings suggest that direct structure/function regression models enhances our understanding of functional connectivity variation, providing insights into underlying mechanisms and discriminability of brain connections.

Keywords: fMRI · Functional Connectivity · Generalized Additive Models · Discriminability · Structural Features · Explainability

D. R. Bathula et al. (Eds.): MLCN 2024, LNCS 15266, pp. 166–175, 2025.
https://doi.org/10.1007/978-3-031-78761-4_16

1   Introduction

Functional connectivity of the brain, which refers to the synchronized activity between spatially distinct brain regions [5], is often measured using functional magnetic resonance imaging (fMRI) and has the goal of understanding network dynamics underlying cognitive processes and neurological disorders. In recent years, there has been a significant interest in studying the relationship between functional connectivity and structural brain features, in particular, how the brain dynamics is regulated by its structural organization and connectivity [2,7,9].

Structural brain features, such as cortical morphology and connectivity density, play pivotal roles in shaping functional connectivity patterns. Recent studies suggest that the degree of functional activation within a cortical area is intricately tied to the physical characteristics of the region, including cortical volume, thickness, surface area, and curvature [13]. Furthermore, the extent of locally exchanged cortical activity appears to be influenced by the density of local connections [3]. Moreover, diffusion imaging and anatomical MRI techniques have been instrumental in providing personalized predictions across a wide spectrum of neurological and psychiatric conditions [10]. Recent research has also explored novel approaches to understanding connectivity variation at the subject level. For instance, [12] characterized functional connectivity at subject level using an edge distribution approach, where they employed the estimated density of connectivity between nodes of interest as a functional covariate, enabling non-geometrically localized connectivity investigations. Similarly, [11] developed regression models on functional connectivity matrices, incorporating structural and regional factors such as geographic distance, homotopic distance, network labels, and region indicators as covariates to explain variations in connections. While these methods explore the relationship between brain structure and function, the precise relationship between spatiotemporal patterns and *local* structural properties remains unexplored. Additionally, these studies are typically conducted under a template or parcellation space, which involves a complex registration process that doesn't adequately account for inter-subject variability. Further research is needed to delve into the specific mechanisms linking local structural features to functional connectivity patterns and to develop methods that better accommodate individual differences in brain structure.

In this work, we introduce a novel and easily implemented *analysis* approach aimed at explaining the variation in functional connectivity by integrating important structural factors such as anatomical morphology summaries, voxel intensity, diffusion-weighted information, and geographic distance. Our methodology utilizes region-pair or vertex-pair information within additive models. We employ subject-level Fisher-transformed connectivity matrices as the outcome of interest and incorporate structural brain factors as covariates to explain the variations in connections. We show that our approach can be performed in template space, as well as subject (vertex) space, thereby accounting for inter-subject differences that get removed via template registration. Additionally, we investigate the discriminability (a measure of repeatability) of data under our proposed approach, which quantifies the probability that two measurements for the same subject will be more similar than measurements for two different subjects. We utilize this measure to analyze the repeatability of brain connectivity measurements between twin pairs and non-twin pairs data available from the Human Connectome Project (HCP). By comparing discriminability in twins, evidence is presented that this analysis method captures underlying connectivity patterns that are repeatable while not removing registration-based inter-subject variation.

2   Methodology

We adopt a statistical approach as shown in Fig. 1 using generalized additive models (GAMs) [8, 16] to capture the relationship between functional connectivity and various structural brain factors. GAMs are particularly advantageous for our analysis due to their non-parametric nature, allowing us to model complex, non-linear relationships between the dependent variable, which is the functional connectivity profile, and a set of independent structural covariates.

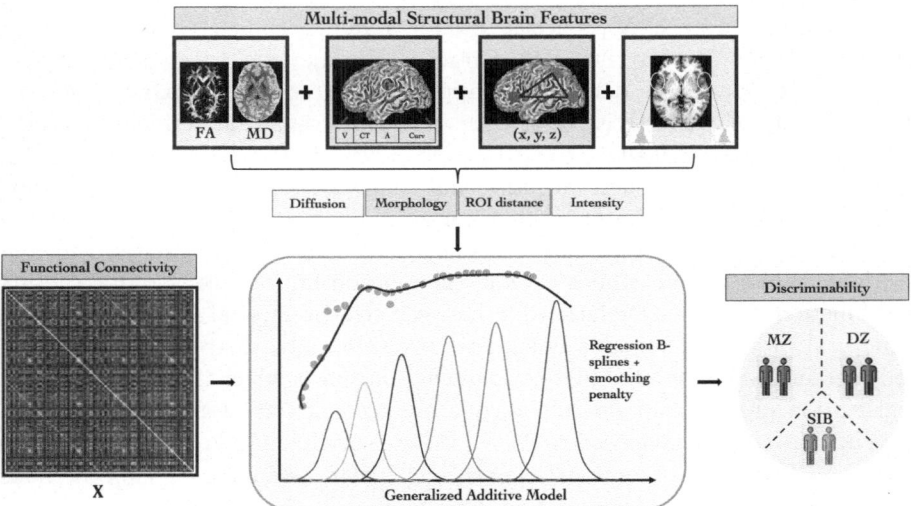

Fig. 1. Integration of multi-modal structural brain features with functional connectivity data for discriminability analysis. The multi-modal structural brain features include fractional anisotropy (FA), mean diffusivity (MD), cortical thickness (CT), gray matter volume (V), surface area (SA), curvature (Curv), combined with region-of-interest (RoI) coordinates (x, y, z) and voxel intensity measures. The outcome of this analysis aims to distinguish between different familial relationships (Monozygotic twins - MZ, Dizygotic twins - DZ, and Siblings - SIB). (Color figure online)

2.1   Connectivity Additive Model

Each subject's functional connectivity profile is described by a matrix of Fisher's Z-transformed empirical Pearson correlations, $\mathbf{X}$, where each element $\mathbf{X}(i,j)$ denotes the correlation between a pair of parcellated brain regions or voxels i and j. To account for the non-localized nature of brain region pairs, we vectorize the upper triangular portion of $\mathbf{X}$. This vector serves as the dependent variable in our additive models.

The independent variables in our regression models include a variety of structural factors. Anatomical morphology summaries are included to capture the shape and size differences between brain regions, voxel intensity metrics to account for the signal characteristics within each region, and diffusion-weighted metrics to assess the integrity of the fiber tracts connecting the regions. Geographic distance, or the spatial proximity between each pair of regions, is also included as a covariate. Our model is formalized as follows: for each subject s, the connectivity between regions i and j is modeled as:

$$\mathbf{X}_s(i,j) = \alpha + f_1(\text{Diffusion}(i,j)) + f_2(\text{Morphology}(i,j))$$
$$+ f_3(\text{Distance}(i,j)) + f_4(\text{Intensity}(i,j)) + \epsilon_s(i,j).$$

Here, α represents the intercept, $f_n(\cdot)$ are smooth functions of the respective structural covariates, and $\epsilon_s(i,j)$ is an error term. The smooth functions, $f_n(\cdot)$, are determined from the data, allowing for the flexibility to capture the unique shape of the relationship between each covariate and the connectivity measure. This is in contrast to traditional linear models that impose a fixed form (linear or polynomial) on the relationship. We fit the smooth terms using penalized regression B-splines and an additive second-derivative smoothing penalty for regularization.

The diffusion-weighted imaging components $f_1(\text{Diffusion}(i,j))$, include mean diffusivity (MD) and fractional anisotropy (FA). MD offers insights into the average rate of water molecule diffusion within brain tissue, indicative of tissue density and cellular integrity. FA measures the directional coherence of water diffusion, reflecting the degree of anisotropy within the white matter tracts, which is crucial for understanding the integrity of neural pathways. For any two regions, we model this term using a cosine similarity metric defined as follows: $\frac{M(i) \cdot M(j)}{\|M(i)\|\|M(j)\|}$ where $M(i)$ and $M(j)$ are vectors consisting of diffusion-weighted imaging descriptors for regions i and j, respectively.

For $f_2(\text{Morphology}(i,j))$, we incorporate smooth terms encompassing four critical structural features-cortical thickness, gray matter volume, surface area, and curvature. Cortical thickness reflects the depth of the cerebral cortex and is associated with cognitive functions. Gray matter volume represents the size of the cortical and subcortical regions, directly linked to brain maturity and development. Surface area contributes to the overall cortical surface, indicating regional developmental patterns, while curvature reflects the folding patterns of the cortex, which may influence neural processing efficiency. We model this similarly using a cosine similarity metric.

The geographical distance term $f_3(\text{Distance}(i,j))$, quantifies the Euclidean distance between brain regions. This measure is fundamental, as geographical correlations could potentially arise from biological reasons such as function specialization, and the spatial distance between regions i and j is computed using the Euclidean distance between their centroid coordinates: $\|C(i) - C(j)\|_2$

Lastly, Voxel intensity $f_4(\text{Intensity}(i,j))$, is modelled through the subject's corresponding T1-weighted image voxel values. This term helps to account for the variation in signal characteristics that might influence functional connectivity measurements. For each region, a histogram $H(i)$ of voxel intensities is plotted. The similarity between histograms for regions i and j is computed using the 2-Wasserstein distance: $\left(\int_{-\infty}^{\infty} |F_{H(i)}(t) - F_{H(j)}(t)|^2 dt\right)^{\frac{1}{2}}$. Here, $F_{H(i)}(t)$ and $F_{H(j)}(t)$ are the cumulative distribution functions (CDFs) corresponding to the histograms of voxel intensities for regions i and j, respectively.

2.2   Data Discriminability

To assess the discriminability between monozygotic (MZ), dizygotic (DZ), non-twin sibling (SIB) and not-related (NR) pairs with respect to functional brain connectivity, we adopt a statistical framework that quantifies the degree to which connectivity patterns are more similar within twin pairs as compared to between non-twin pairs. This approach extends the concept of discriminability introduced by [1,15] to the domain of data repeatability. The discriminability statistic, δ, is defined as the probability that a randomly chosen within-pair distance is less than a between-pair distance, under a given distance metric $d(.,.)$. Higher values of δ indicate greater repeatability. The within-pair and between-pair distances are calculated using the chosen metric between the measurement profiles of the subjects in the pairs. Mathematically, the unbiased estimator of discriminability for each group can be expressed as:

$$\delta_{\text{group}} = \frac{1}{P(P-1)T^2(T-1)} \sum_{p=1}^{P}\sum_{\substack{q=1 \\ q \neq p}}^{P}\sum_{i=1}^{T}\sum_{j=1}^{T} \mathbb{I}\{d(Z_{p,i}, Z_{p,j}) < d(Z_{p,i}, Z_{q,j})\}$$

where P is the number of pairs in the MZ, DZ, SIB or NR group, T is the number of twins (or siblings) per pair ($T = 2$), $Z_{p,.}$ represent the vector of measurement values for a pair p and $\mathbb{I}$ is the indicator function which evaluates to 1 when the condition within is true, and 0 otherwise. We chose 60 pairs from each of the MZ, DZ, SIB, and NR groups within the HCP dataset for conducting the discriminability analysis because they included all the imaging modalities, and we employed Euclidean distance as the metric for measuring distances.

3   Experiments

3.1   Human Connectome Project Data

The dataset for our study comprises resting-state and task fMRI data obtained from the Human Connectome Project (HCP) [14], specifically from the HCP

900 subject release where imaging was conducted using a Siemens Skyra 3T scanner at Washington University in St. Louis. In this study, we exclusively utilized data from the left-to-right phase encoding sessions to ensure consistency and minimize potential directional biases in the analysis. For parcellation-based analysis, we employed the Destrieux atlas [4] that consists of 148 brain ROI's. The preprocessing protocol was adapted from the HCP "fMRIVolume" pipeline [6], which incorporates a series of steps such as gradient unwarping, motion correction, distortion correction with FSL's topup tool, registration to structural T1-weighted scan, non-linear registration into MNI152 space, grand-mean intensity normalization, and spatial smoothing using a Gaussian kernel with a full-width half-maximum of 4 mm. Crucially, while coregistration to structural T1-weighted scans was carried out, we intentionally omitted non-linear registration to standard space (MNI atlas) for the subject space analysis to preserve the unique anatomical features of each subject. Subsequently, we extracted time series data for each parcel defined by the Destrieux atlas for 148 regions. For subject space (vertex level) analysis, we perform a spatial stratified sampling defined by the atlas in subject space, where we sample 5000 vertices in total. The Fisher's Z-transformation was applied to the correlation coefficients, resulting in $\binom{148}{2}$ correlations for atlas space analysis, and $\binom{5000}{2}$ resulting in approximately 12 million pairs, to normalize the distribution of connectivity measures. This transformation attempts to ensure that the resulting scores approximate a normal distribution. The normality of Z-scores is useful for both statistical inference and general performance of the GAM, which used a squared error loss function (in our case) and thus results in a more accurate error model for our data. It should be noted, however, that the GAM approach could be used for binary data if threshold using a Bernoulli log-likelihood and GAM link function.

3.2   Partial Dependence Analysis

To understand the influence of each structural feature independently, we perform partial dependence analysis. This technique allows us to visualize the effect of a single structural covariate on the predicted outcome of functional connectivity while averaging out the effects of other covariates. For a given covariate k, the partial dependence function PD is defined as: $PD(X_k) = E_{X_{\sim k}}[f(X_k, X_{\sim k})]$ where $X_{\sim k}$ represents all features except X_k and f is the predictive model.

3.3   Implementation Details

For our analysis, we employed the pyGAM framework. Given that the data follows a normal distribution, which is part of the exponential family, we selected the identity link function for this model. To capture the nonlinear patterns in the data, we utilized a third-order spline term, incorporating 20 splines to fit the model effectively. Further, to prevent overfitting and ensure smoothness, a second derivative smoothing penalty was applied to the spline terms in the model. The optimal smoothing parameter was determined through a comprehensive grid search, with $\lambda = 0.6$ chosen to balance the trade-off between fit and smoothness.

4   Results and Discussion

We used explained deviance as our measure of fit, with our model explaining 15.7% of the variability in the response variable. Figure 2 represents the partial dependence plots of different structural features against functional connectivity values for resting state in template and vertex space. The diffusion-weighted metrics show a consistent pattern across both spaces, with functional connectivity values remaining relatively stable, regardless of the variation in cosine similarity. This suggests that raw diffusion properties, captured by metrics like MD and FA, may not vary significantly with functional connectivity, or may be influenced by other unmeasured factors. Of note, these are not measures of diffusion based connectivity between the pairs; instead only measuring the distance in the intra-regional diffusion properties.

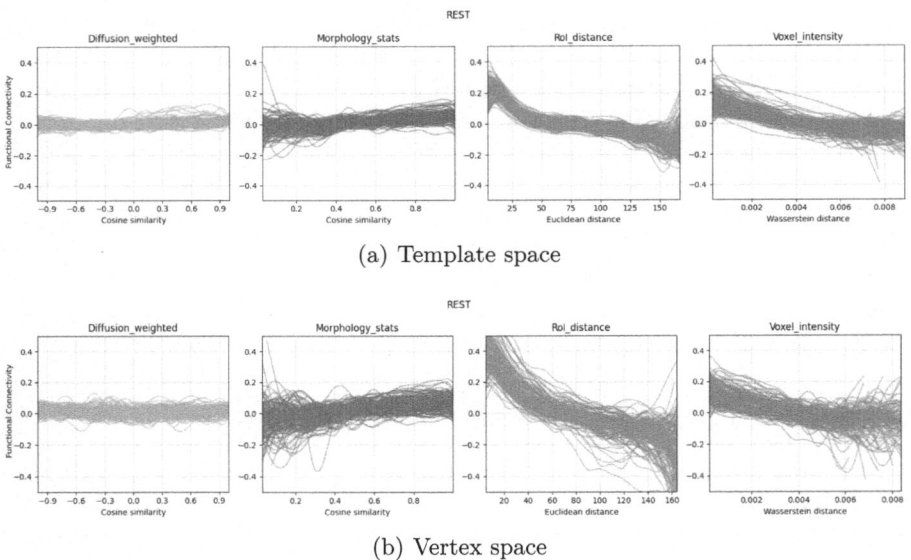

(a) Template space

(b) Vertex space

Fig. 2. Partial dependence plots of different structural features against functional connectivity values for resting state in **Top:** template and **Bottom:** vertex (subject) space. Each figure contains four panels that correspond to diffusion-weighted metrics, anatomical morphology summaries, region of interest (RoI) distance, and voxel intensity.

The non-diffusion structural summaries reveal a more pronounced effect on functional connectivity. During the resting state, there is a noticeable trend where higher morphological alignment (higher cosine similarity) correlates with increased functional connectivity. Interestingly, this trend appears to be consistent across all tasks (provided in the supplementary material), indicating a robust morphological influence on connectivity. The ROI distance panels exhibit an expected inverse relationship; as the Euclidean distance between ROIs increases, functional connectivity decreases. This inverse relationship is consistent across both spaces, where it aligns with the intuitive notion that closer brain regions are more likely to exhibit stronger connectivity. Voxel intensity, represented by the Wasserstein distance of T1-weighted image voxel values, shows a distinct negative correlation with functional connectivity in all cognitive states. The strength of this correlation seems to be most pronounced in the resting state, with a steady decline in functional connectivity as voxel intensity dissimilarity increases. The consistency of these patterns across different cognitive states suggests that these structural features have a stable relationship with functional connectivity, independent of the cognitive demands placed on the individual.

Table 1. Discriminability analysis comparing MZ, DZ, SIB, and NR individuals across various cognitive states in template and vertex spaces. Vertex space results are visualized through a gradient of grey color from darker to lighter, moving from MZ to NR, where a darker color denotes higher degree of repeatability.

Level	Cognitive state	f_2(Morphology)				f_1(Diffusion-weighted)				GAM predictions			
		MZ	DZ	SIB	NR	MZ	DZ	SIB	NR	MZ	DZ	SIB	NR
TEMPLATE	Rest	0.58	0.58	0.51	0.47	0.71	0.59	0.53	0.50	0.58	0.55	0.49	0.49
	WM	0.52	0.56	0.50	0.51	0.64	0.62	0.53	0.45	0.55	0.50	0.48	0.50
	Relational	0.51	0.59	0.45	0.46	0.61	0.53	0.45	0.38	0.51	0.51	0.47	0.42
	Language	0.50	0.52	0.50	0.48	0.60	0.63	0.47	0.44	0.60	0.50	0.46	0.47
	Emotion	0.48	0.57	0.48	0.46	0.61	0.56	0.49	0.44	0.55	0.52	0.51	0.47
	Gambling	0.50	0.45	0.49	0.47	0.68	0.53	0.55	0.44	0.55	0.48	0.49	0.45
	Motor	0.53	0.49	0.51	0.51	0.66	0.55	0.49	0.45	0.56	0.53	0.51	0.49
	Social	0.53	0.60	0.49	0.50	0.64	0.51	0.50	0.46	0.56	0.48	0.54	0.44
VERTEX	Rest	0.50	0.51	0.32	0.29	0.64	0.59	0.36	0.30	0.53	0.54	0.33	0.29
	WM	0.49	0.56	0.34	0.37	0.66	0.61	0.34	0.32	0.54	0.50	0.34	0.34
	Relational	0.49	0.52	0.35	0.28	0.63	0.52	0.35	0.30	0.50	0.51	0.35	0.30
	Language	0.52	0.50	0.33	0.32	0.62	0.57	0.38	0.33	0.59	0.50	0.33	0.35
	Emotion	0.44	0.49	0.34	0.32	0.60	0.52	0.35	0.30	0.54	0.52	0.36	0.32
	Gambling	0.51	0.52	0.35	0.33	0.64	0.51	0.38	0.33	0.55	0.47	0.34	0.31
	Motor	0.49	0.50	0.34	0.36	0.65	0.51	0.37	0.35	0.54	0.53	0.36	0.36
	Social	0.49	0.58	0.37	0.32	0.64	0.52	0.38	0.35	0.55	0.48	0.38	0.29

Table 1 presents the discriminability (δ_{group}) analysis outcomes for MZ, DZ, SIB, and NR individuals across various cognitive states in both template and vertex spaces using partial dependence functions from morphology, diffusion, and GAM predictions. We used the spline coefficients of each feature from partial dependence analysis to compute the distance (d) in discriminability analysis, thereby combining PD analysis to derive the δ_{group} scores shown in Table 1.

This approach ensures that the influence of each structural feature on functional connectivity is reflected in the discriminability scores, providing an integrated evaluation of our framework. In template space, the δ values for MZ, DZ, and SIB are slightly higher than NR, suggesting genetic influences, but the differences are minimal, around the 0.50 mark, indicating limited discriminability. Vertex-level analysis, however, consistently shows higher discriminability and clearer distinctions between twins and non-twin pairs across all cognitive states, especially with diffusion-weighted information. We provide bootstrapped standard errors (SE) for δ to account for uncertainties due to structural measurement errors and environmental factors. For instance, at the vertex level (during rest), the SEs for the Morphology function for MZ/DZ/SIB/NR is 0.029/0.036/0.034/0.047, and for the GAM predictions is 0.033/0.036/0.044/0.049, respectively. Despite these uncertainties, the ranking (MZ>DZ>SIB>NR) remains consistent for diffusion-weighted factors, highlighting their utility as biomarkers for understanding familial relationships. Vertex level analysis shows to be a more consistent method for assessing familial relationship between twins and non-twin pairs, highlighting its advantage over template space analysis.

5   Conclusion

We have developed a novel analytical framework that integrates an array of structural determinants, encompassing anatomical morphology summaries, voxel intensity metrics, diffusion-weighted measures, and geospatial distances, offering a multi-modal analysis of the impact of these structural elements on the dynamics of functional brain networks. Our approach utilizes subject-specific connectivity matrices and is adaptable to both template and individual-specific spatial domains, introducing a unique way for accommodating inter-individual variability-a crucial aspect frequently neglected in traditional analyses. The employment of our framework in analyzing HCP data, particularly for the evaluation of connectivity heritability in twin cohorts, highlights our framework's ability to understand the inheritable components of brain connectivity patterns. The results of discriminability assessments further confirms the robustness and reproducibility of our technique, highlighting its efficacy in discriminating biological variances within brain connectivity configurations.

Acknowledgements. This work was supported by the National Institutes of Health grant R01 EB029977 (PI Caffo) from the National Institute of Biomedical Imaging and Bioengineering and the National Institutes of Health grant R01 HD108790 (PI Venkataraman) from the National Institute of Child Health and Human Development.

Disclosure of Interests. The authors declare that they have no competing interests in the paper.

References

1. Bridgeford, E.W., et al.: Eliminating accidental deviations to minimize generalization error and maximize replicability: applications in connectomics and genomics. PLoS Comput. Biol. **17**(9), e1009279 (2021)
2. Bullmore, E., Sporns, O.: Complex brain networks: graph theoretical analysis of structural and functional systems. Nat. Rev. Neurosci. **10**(3), 186–198 (2009)
3. Bullmore, E., Sporns, O.: The economy of brain network organization. Nat. Rev. Neurosci. **13**(5), 336–349 (2012)
4. Destrieux, C., Fischl, B., Dale, A., Halgren, E.: Automatic parcellation of human cortical gyri and sulci using standard anatomical nomenclature. Neuroimage **53**(1), 1–15 (2010)
5. Friston, K.J.: Functional and effective connectivity: a review. Brain Connectivity **1**(1), 13–36 (2011)
6. Glasser, M.F., et al.: The minimal preprocessing pipelines for the human connectome project. Neuroimage **80**, 105–124 (2013)
7. Guye, M., Bettus, G., Bartolomei, F., Cozzone, P.J.: Graph theoretical analysis of structural and functional connectivity MRI in normal and pathological brain networks. Magn. Reson. Mater. Phys., Biol. Med. **23**, 409–421 (2010)
8. Hastie, T.J.: Generalized additive models. In: Statistical Models in S, pp. 249–307. Routledge (2017)
9. Honey, C.J., et al.: Predicting human resting-state functional connectivity from structural connectivity. Proc. Natl. Acad. Sci. **106**(6), 2035–2040 (2009)
10. Sabuncu, M.R., Konukoglu, E., Initiative, A.D.N.: Clinical prediction from structural brain MRI scans: a large-scale empirical study. Neuroinformatics **13**, 31–46 (2015)
11. Smith, B., Zhao, Y., Lindquist, M., Caffo, B.: Regression models for partially localized fMRI connectivity analyses. Front. Neuroimaging **2**, 1178359 (2023). https://doi.org/10.3389/fnimg.2023.1178359
12. Tang, B., et al.: Differences in functional connectivity distribution after transcranial direct-current stimulation: a connectivity density point of view. Hum. Brain Mapp. **44**(1), 170–185 (2023)
13. Tillisch, K., et al.: Brain structure and response to emotional stimuli as related to gut microbial profiles in healthy women. Psychosom. Med. **79**(8), 905–913 (2017)
14. Van Essen, D.C., et al.: The WU-Minn human connectome project: an overview. Neuroimage **80**, 62–79 (2013)
15. Wang, Z., Bridgeford, E., Wang, S., Vogelstein, J.T., Caffo, B.: Statistical analysis of data repeatability measures. arXiv preprint arXiv:2005.11911 (2020)
16. Wood, S.N.: Fast stable restricted maximum likelihood and marginal likelihood estimation of semiparametric generalized linear models. J. R. Stat. Soc. Ser. B Stat Methodol. **73**(1), 3–36 (2011)

Author Index